Research into the Area of Supply Chain

Albert Corominas

OmniaScience

Open Access Support

If you find this book interesting, we would appreciate that you supported its author and OmniaScience so that books can continue publishing in Open Access.

You can contribute at the following link: https://doi.org/10.3926/oss.36

Research into the Area of Supply Chain

Author: Albert Corominas
Universitat Politècnica de Catalunya

Original title: *La Investigació en l'àmbit de la Cadena de Subministrament* edited by OmniaScience 2017
Translation by Martin Moore (moorcamp@orange.es), revised by author

ISBN: 978-84-946352-8-1
DOI: https://doi.org/10.3926/oss.36

ACKNOWLEDGMENT

The author acknowledges the support of the Supply Chain and Operations Management Division of the Institute of Industrial and Control Engineering (Universitat Politècnica de Catalunya) to make possible this edition.

INDEX

NOTES

Note 1. About the origin of this text, industrial engineering and its relationship with the supply chain

This work has its origins in the lectures taught by the author, between 2007 and 2017, in the subjects "Introduction to logistics research", from the Master's degree in Logistics, Transport and Mobility (from 2014-2015, "Introduction to supply chain research", from the master's degree in Supply Chain, Transport and Mobility) and "Introduction to industrial engineering research", from the master's degree in Industrial Engineering. Therefore, it can be used, and in part it is designed for this purpose, as material for these subjects or similar.

Industrial engineering is an expression that has a long tradition, but its meaning has changed gradually over the years[1].

---

[1] https://en.wikipedia.org/wiki/Industrial_engineering:

> "While originally applied to manufacturing, the use of 'industrial' in 'industrial engineering' can be somewhat misleading, since it has grown to encompass any methodical or quantitative approach to optimizing how a process, system, or organization operates. Some engineering departments and universities have changed the term 'industrial' to broader terms such as Industrial and Manufacturing Engineering, Industrial and Systems Engineering, Industrial Engineering & Operations Research, Industrial Engineering & Management."

As indicated in section 2.1 of this text, what was known until 2016 as the Institute of Industrial Engineering (IIE) is now called the Institute of Industrial and Systems Engineering (IISE).

Currently, in my opinion, industrial engineering is focused primarily on the supply chain, as we shall see.

Given the great diversity of themes in this field, it is common, depending on who is talking and according to their interests and experience, for emphasis to be laid on one aspect or another. Therefore, in spite of all efforts to avoid it, this text is quite probably not free of this kind of bias.

Note 2. On the objectives of this text

The student who wants to do, or is considering the possibility of doing, a PhD or, in general, people who wish to undertake research activities are often faced with the difficulty of knowing what research is, how and in what environment it is carried out and how to present the results. The answers to these questions are not well-known to non-researchers and they usually learn more or less slowly, through contact with research groups, often in the process of preparing a doctoral thesis.

Also, it is extremely important that people who want to make innovative contributions to an organisation know how to distinguish research from non-research, and where to find the results.

In the area of the supply chain there are two added difficulties. On the one hand, the lack of a generally accepted definition of the concept itself, compounded by the little semantic correspondence between the term "supply chain" and the content of the concept referred to. Additionally, in the area of the supply chain, as in all management, apart from the research publications there is a proliferation of others: informative, speculative or even what could be classified as self-help, all of which are at times presented as if they were works of research.

Consequently, this lengthens the time elapsed between the moment that the student decides to go in for research and the point when s/he really begins to do it, as well as affecting productivity and the quality of the research activity.

There are few publications that help students in overcoming these difficulties. In this text I quote some, but the aim is that the text itself will serve for the student to enter quickly into the world of research and, in particular, into research in the field of the supply chain.

Any consideration relating to the philosophy of science and its different currents have been excluded[2]. Similarly, a general discussion about the scientific method has also been excluded, since in the field of the supply chain some very different problems and issues co-exist and must be addressed by a variety of methodologies.

Note 3. Facts and opinions

In this text facts are described and opinions are given. With regard to the latter, if there is no explicit reference they are those of the author, who is solely responsible for them. As a result, some portions are written in first person singular, a rare thing in teaching or research texts, but which is clearer on certain occasions.

Finally, I would like to express my gratitude to Professor J. Olivella for the careful reading of a previous version of this document and the suggestions that, without a doubt, have helped to improve it. However, I am solely responsible for the views expressed here, as I have already indicated, as well as for any errors that may have occurred.

[2] With the exception of the comments of the footnote 35, page 51.

CHAPTER 1

WHAT IS RESEARCH, WHERE IS IT CARRIED OUT AND WHO CAN AVAIL OF IT?

According to the *Frascati Manual*[3], "research and experimental development (R&D) comprises creative and systematic work undertaken in order to increase the stock of knowledge – including knowledge of humankind, culture and society – and to devise new applications of available knowledge".

Therefore, the purpose of research is to increase knowledge (finding new applications for available knowledge is also a way of increasing that knowledge). Now, we can ask what knowledge do we mean when we talk about increasing the knowledge. When someone studies a book it increases *their* knowledge, but of course they are not doing research. However, sometimes searches via the Internet or field work can be considered as research (someone, a primary school student, for example, may not know the names of the streets around his/her school and discovers them by means of a systematic walk round the area). How-

[3] OECD, *Frascati Manual 2015 Guidelines for Collecting and Reporting Data on Research and Experimental Development*. Retrieved October 2, 2017 from http://www.oecd-ilibrary.org/science-and-technology/frascati-manual-2015_9789264239012-en

ever, these activities do not reveal anything that is not already widely known and is within reach of anyone with the training needed to understand it. Certainly, this increases the knowledge of the persons involved, and this can in some cases lead them to think that they have discovered something that nobody knew. Therefore, in this case we could speak of *subjective research*. The person who does it might believe that it is research, but it is not, because only *objective research* can be considered properly as research; that is to say, activities that are intended to generate new knowledge (whether they are successful in doing so or not), in the sense that, although it may seem overly solemn or rhetorical, it is new to humanity[4]. This and nothing else is what is meant when it is said, straightforwardly, that the result of a research activity must be original[5,6].

Then, it is clear that before undertaking research it is necessary to determine very precisely what is already known about the subject concerned[7]. Therefore, to conduct research and present the results it is necessary to establish the corresponding *state of the art*. This requires considerable effort since, although there are now very powerful search and consultation tools available, the volume of publications on any subject, even it is much specialised, is immense. But it is an unavoidable effort, because it reduces the danger of reinventing the wheel and because without a solvent state of the art no research paper is publishable in any serious journal

---

[4] Many research papers are published, but very few that reflect upon research and the people who carry it out (although lately there has been an increase, possibly as a result of the growing volume of research activity throughout the world). Among these: Booth *et al.* (2008), Nielsen (2004), Creedy (2008), Silver (2009), Sodhi and Tang (2014). A reading of these texts is especially recommended for the PhD student who is starting to research, and everybody who plans to dedicate themselves to it professionally. Those already involved will find interesting items, in particular on the publication process, in Clark *et al.* (2016).

[5] I disagree radically, therefore, with Booth *et al.* (2008) when, as a response to the question "What is research?" they claim that "in the broadest terms, we do research whenever we gather information to answer a question that solves a problem" and give as examples of problems, finding a spare part for the car, knowing when Michael Jordan was born or obtaining more information about a new species of fish.

[6] Of course, originality does not imply importance.

[7] Here it is necessary to specify that it should be considered that humanity has knowledge when this is within the reach of everyone who can understand it, that is to say when it has been published in some reasonably accessible support. If knowledge is reserved for a person, a group of people or an organisation, obtaining it by other means and making it public should also be considered a form of research.

(indeed, there are some journals which are not: this topic will be discussed later). It is surprising, therefore, that the introductory texts to research do not stress the importance of preparing the state of the art, or even do not mention it.

However, can every activity that generates new knowledge for mankind be considered as research? An extreme example: counting how many leaves there are on a particular tree; it is a laborious and difficult task, the result of which is of no apparent use or interest to anyone. At another extreme, from the point of view of usefulness, we could show many examples, such as how to establish an effective procedure for carrying out nuclear fusion. But in between these two extremes there are innumerable cases whose status as research is somewhat dubious, such as making a national list of the companies that have a department of quality management, for instance.

The previous paragraph may suggest, but in no case affirms, that counting the leaves on a tree is not research and that setting up an effective process of nuclear fusion is. But, why is leaf counting not research? Because it is useless? No, there are many discoveries that apparently do not serve any purpose and really do end up being useless, yet they are considered as research. Because the result is of no interest to anyone? Certainly, if nobody is interested in the result it is very doubtful that the activity that obtained it may be considered as research. But, how many people should take an interest in the result of a research for it to be recognised as such? So, why is the hypothetical leaf count not research? Because the result is an isolated piece of information, which provides no elements for either building a theory or validating it; it is neither a very valuable nor essential tool for use in other research (as in their day were, for example, the establishment of the universal gravitational constant or of the speed of light in a vacuum).

Later (section 4) it is discussed what the results of research can be, but I will say here that I do not think that mere data collection can be considered as research if it is not carried out within the framework of the development and validation of a theory, that is, a set of laws that serve to relate a specific order of phenomena, a model of reality that allows you to obtain the answers about their behavior[8].

---

[8] Steven Toms, in Chapter 29 ("Publishing historical papers in management journals and in business history journals") of Clark *et al.* (2016) says "Where the researcher, for example, notes a previously undiscovered fact about fish processing in the nineteenth century, unless they can also show how the wider literature should be modified in consequence, the research will be unsuitable for publication" (pp. 248-249).

According to this, counting the leaves on a tree or making a list of the companies that have a department of quality management may or may not be research, depending on the context in which these activities take place. For instance, counting the leaves could be a contribution to a theory on climate change.

I think the reflections of Robert Hutchins, in his book *The University of Utopia* (Hutchins, 1953) are appropriate in this case. Hutchins says that "the accumulation of data about some subject, even a subject of great importance" does not fall to the university, "unless they are able to think and communicate about the ideas involved in these phenomena". Because, according to Hutchins, "research is thinking about important problems" and "if research does not involve thinking, as I believe a great deal that is called research does not, then it has no place in a university". "Knowledge is organized information, i.e. information that has been reflected upon, thought about. The collection of information for the purpose of thinking about it is a legitimate function of a member of a university, but only on the assumption that he goes on to think about what he collects" as Hutchins also says.

And nuclear fusion? In the current state of knowledge a lot of research is still needed, of course, but to achieve the industrial exploitation of the phenomenon, technological development and innovation are also necessary. These three concepts, research (R), development (D) and innovation (I) are often represented as the three successive parts of a whole: **R** + **D** + **I** (in bold in this paragraph to avoid typographic confusion), although this sequence does not always occur: the **R** can stand alone, there may be **R** & **D** without **I** and there may also be **I** without **R** & **D** (on the other hand, the **D** cannot stand alone: there must be an **R** before it). The **R** corresponds to the research, the contribution of new knowledge. The **D**, development, allows the knowledge to be passed on to the prototype. Innovation, **I**, refers to the production processes and the products that can be derived from **R** & **D** or, quite simply, is the result of the idea of incorporating, for example, a new function in a product or applying a method known to improve a process.

Therefore, research is done in:

- Public research bodies.
- Universities (in Catalonia, as in all of Spain and in most European countries, essentially, but not exclusively, in the public universities; in the United

States, on the other hand, public and private are represented more evenly among the universities that carry out research). In Catalonia and the rest of Spain, the basic units of research are the research groups, which can work in the university research institutes or departments.

- Companies[9] and associations of companies (in fact, the research itself is carried out only in a few large, or very large, companies and, more rarely, in organisations run by consortia of companies, with or without public sector participation).

The development, in technology centres, engineering units from universities and in the companies.

Innovation only occurs in companies, because it is where the production processes take place. Universities can be innovative with regard to their own activities, but, as such, universities cannot innovate products or processes in the automotive sector, for instance, even though the results of the research carried out can lead to innovation in that sector and can create innovative companies that take advantage of R & D results obtained at the university itself[10].

---

[9] In this text, so as not to encumber the writing, the term company, if it is not explicit or inferring something else from the context, refers to any organisation that carries out a process of production of goods or of services. It may be, therefore, a public or private company, public administration or very different types of organisations that are not businesses, such as associations of all kinds, sports clubs or NGOs.

[10] In Aghion *et al.* (2008) it is said: "if by being entrepreneurial it is meant that universities should be attuned to their environment, both social and economic, and reactive to it, then we agree. But if it means that universities should become very active agents in the market for 'innovation' and that they should strive to obtain a significant amount of financing in this way, then we are more reluctant. There may be better-designed institutions for this, for example technological centres and technological parks located close to universities (even with the participation of the latter in their management). Raising money through direct entrepreneurial activities may be tempting (much of the needed equipment and human resources being in place and, perhaps, paid for) but the quantitative significance of these funds may easily be overestimated. The university has a core mission that is not business. It is education and research that only universities – and research centres – can accomplish: what is now commonly called 'frontier research'. University research is heavily subsidised because it is, or it should be, of the long-term, high-risk variety that could not be developed in the marketplace".

It is not absolutely impossible to do research individually and without planning, but almost all the relevant results are obtained by research teams working on projects planned for a timeframe of several years.

The funding of the projects, in addition to what they implicitly receive for being carried out in an institution, is normally obtained from calls for proposals based on research plans that may be at a state (*Plan Nacional de Investigación* in Spain) or international level (in our case, mainly from the European Union).

Another question is who is interested in the research or, more precisely, the results of the research: potentially, all the organisations involved in production processes. Although there may be innovation without research, there is no doubt that a fundamental part of innovation is derived from research (just think about the developments in information technology and communications in recent decades) and that if a company wants to be consistently innovative it should strive to be aware of the results of research.

Research concerns, obviously, the student who wants to do it, who wants a doctorate and wants to pursue a career in the field. And there is also the student who wants to develop professionally in companies, because if you want to contribute to innovation you must keep up to date with regard to the results of research and must know how to identify what is relevant to your goals.

Before I get into the field of the SC, we return to Hutchins (1953): "In the educational system, and particularly in the university, there is the problem of forming a thinking community when the members of the community cannot think together because they cannot communicate with one another. Specialisation means that specialised men cannot think together because their training and their work have split them off from other men"; "Society requires specialists; but even specialism requires, if it is not to come to a dead end, that every specialty be able to throw light on every other. Every specialist must therefore be able to catch whatever light is being thrown from any quarter". Thus wrote Hutchins more than sixty years ago and since then the fragmentation of knowledge has progressed rapidly. The scientific community is increasingly segmented into groups specialising in very specific themes and which generally do not pay much attention, if any, to developments taking place in other fields. This state of affairs undoubtedly fevers the acquisition of numerous incremental results, but does not bring about any leaps in the advancement of knowledge. On the other hand, the windows opening to other areas permit the learning of methods and techniques that are new to the area and which favour renewal and progress (think, for example, of the renewal and the momentum that the adoption of techniques from artificial intelligence brought about in operations research). Therefore, it is advisable that, as Hutchins said, the researcher "be able to catch whatever light is being thrown from any quarter" and tries to find it. Along the same lines Herbert Simon (Singhal and Singhal, 2012b) advised, with regard to a PhD, "to stimulate the imagination, arm yourself with knowledge from many fields in order to approach the task from different angles".

The rest of this document is organised as follows. Section 2 is devoted to the thematic and terminological evolution that has resulted in what is now the area of the supply chain. Section 3 focuses on the network of organisations, activities, and publications that make up the world of research in our field. The types of subjects for a thesis or a research article are dealt with in Section 4. Section 5 describes the process of publishing an article. And the text concludes with a diagnosis of the current situation and a proposal of perspectives.

CHAPTER 2

TERMINOLOGICAL AND THEMATIC DEVELOPMENTS IN WHAT IS NOW THE AREA OF THE SUPPLY CHAIN

2.1. Elements of the history of the discipline

Knowing the history of a discipline allows you to better understand the present and helps to prevent relapses in errors or previous discoveries.

This present results from the confluence of several streams which originally developed autonomously. On the one hand, the development of production systems and the management of operations within it. On the other, the use of quantitative methods, which have received a decisive boost from operations research and computer science.

The concept of supply chain is the result of a broadening of our point of view of the processes of supply, production, distribution and recovery, which have been taking place since the dawn of humanity. Even the collection of rare materials, valuable for use or exchange, (flint, gold, copper, etc.) goes back to the most primitive times.

The history of these processes is poorly documented until we arrive to Adam Smith (1723-1790)[11,12] and, in particular, his work *An Inquiry into the Nature and Causes of the Wealth of Nations* (1776), in which, just at the beginning, he described production in a pin factory, in which the process had been divided into 18 operations, distributed among 10 workplaces. Beyond the description, Smith does highlight that this division of labour leads to a considerable productivity increase and attributed this to the ease of learning elementary operations and the skill acquired by repeating them, saving time in changing tools and also, although this was not yet a reality but a forecast of Smith, the possibility of automation brought about by the division of the process into elementary operations[13].

Charles Babbage (1791-1871), probably best known as a pioneer of computer science, is of interest here as a founder of industrial engineering as a scientific-technical discipline. In fact, he was the first to say that apparently dissimilar production systems had common principles, which he had deduced from his own professional experience. Babbage added quantification (time and cost of each operation) to Adam Smith's description of the pin factory as well as an advantage that Smith had not mentioned, which is that as the level of remuneration of each person corresponds to the assigned task requiring the highest qualifications, the division of labour implied a reduction in the cost of the whole process.

Babbage was ahead of his time. The contributions of Frederick Winslow Taylor (1856-1915), on the other hand, came when the production systems, in countries such as the United States, had developed in such a way that they could adopt and take advantage of them. The triumph of Taylorism was possible because the division of labour was strongly established in a significant part of the pro-

---

[11] There are few exceptions. The most notable are the descriptions of the operation of the Arsenal of Venice. In this regard, see George Jr. (1968) and http://www.cabovolo.com/2008/02/el-arsenal-de-venecia-la-primera-fbrica.html, http://ca.wikipedia.org/wiki/Arsenal_de_Ven%C3%A8cia.

[12] You can find more information about these authors and their works on the Internet and, especially, in Wikipedia.

[13] A surprising precedent of Adam Smith, with regard to its description of the advantages of division of labour and on the consideration that this is possible when the market is big enough, is located in the *Cyropaedia (The Education of Cyrus)*, of Xenophon, written around 370 BC (I am grateful to Dr. Alberto García-Villoria for having given me this information).

duction systems and also because René Descartes' (1596-1650) ideas about the scientific method had been applied: to be methodical, doubt systematically, analyse, synthesise and make use of lists (now we would say *check-lists*). In short, Taylor measures time and analyses the methods in order to establish the best way to do a job, proposing an approach to work characterised by fragmentation, regulations (the person who performs the task has to do so according to the stipulated methods) and individualisation (the object of study is the work place and not the production system).

Frank Bunker Gilbreth (1868-1924) studied the movements of work in greater depth, making use of cinematographic films and defining a list of basic movements (*therbligs*) with which you can describe any work process. In this way, he established the bases of the predetermined time systems. On the other hand, he applied the study of work to activities where it had never been done before: construction, health care and the military. He worked together with Lillian Evelyn Moller (1878-1972, also known as Lillian Moller Gilbreth), an industrial psychologist, forming a couple who had twelve children[14]. During the forty-eight years after Frank's death, Lillian continued making important contributions to industrial organisation.

Although the moving assembly line has often been attributed to Henry Ford (1863-1947), its origins can be traced back to the Arsenal of Venice; the system of quartering animals in the slaughterhouses of Chicago can also be considered a precedent which is said to have inspired Ford. In any case, Ford had the great merit of revolutionising the automobile industry and converting the private car from a luxury product into an object within reach of a very large proportion of the population. It was possible due to the design of the product (the Ford Model T), the assembly line production based on the highly advanced division of labour and, above all, and this is what is of most interest here, for its integrated conception of the production system and logistics. It can now be said that Ford was the first to conceive, implement and put into practice a great supply chain.

Walter Rathenau (1867-1922), German entrepreneur and politician, introduced the concept of aggregate planning.

---

[14] https://en.wikipedia.org/wiki/Cheaper_by_the_Dozen and https://en.wikipedia.org/wiki/Cheaper_by_the_Dozen_(1950_film).

In 1916 Henry Fayol (1841-1925) published *Administration industrielle et générale*, the foundational work in terms of business administration and the role of management.

In parallel to these industrial and conceptual contributions, a quantitative line of solving management problems with the support of mathematical models was developed. In 1909, Agner Krarup Erlang published the article that started the queueing theory, based on the study that had been carried out on the traffic of calls to the Copenhagen telephone exchange. The formula to calculate the EOQ (*Economic Order Quantity*), the work of Ford Whitman Harris (1877-1962), was published in 1913.

However, quantitative techniques did not upsurge until the beginning of World War II (1939-1945) when the British army involved teams of scientists to analyse and improve military operations. This led to *operational research* in the United Kingdom and *operations research* in the United States, that is to say the research into military operations that, given the results achieved in the war, was later applied to the civilian field. The operations research arose as a result of the contribution of numerous scientists, grouped together in multidisciplinary teams. Of these, the most famous was that known as the *Blackett Circus*, led by the physicist (awarded the Nobel Prize in 1948) Patrick Maynard Stuart Blackett (1897-1974). Philip McCord Morse (1903-1985), regarded as the founder of operations research in the United States, was also a physicist.

In 1946, ENIAC (*Electronic Numerical Integrator and Computer*) was presented in public, the first programmable and general application electronic computer. It was a unique moment in the constantly developing process of calculation methods, which has decisively expanded the possibilities of applying quantitative techniques. A few years later the first commercial computers went on sale; the BINAC (1949, not a success) and the UNIVAC I (*UNIVersal Automatic Computer* I, 1951).

The simplex algorithm, for solving linear programs, was presented by George B. Dantzig (1914-2005) in 1947.

In 1948 the *American Institute of Industrial Engineers* was founded (from 1981, *Institute of Industrial Engineers*, and since 2016, *Institute of Industrial and Systems Engineers*: IISE), in the United States, and *the Operational Research Club* (later, from 1953, *Operational Research Society*: ORS), in the United Kingdom.

The IISE is a predominantly professional association, but it also publishes scientific journals.

The ORS is the world's oldest scientific association of operational research. Its activity reflects a very broad concept of operational research oriented towards application, in accordance with the historical origins of the discipline.

The ORSA (*Operations Research Society of America*) was founded in the United States in 1952 and the following year, the TIMS (*The Institute of Management Science*) was established. Both societies had some overlapping thematic areas and drew closer with time, until they merged in 1995 under the name of INFORMS (*Institute for Operations Research and Management Science*). In fact, OR and MS, in one order or another, are terms that often go together (for example, they constitute a unique category in the main journal databases).

Additionally, also in the United States, the APICS (*American Production and Inventory Control Society*) was founded (1957), with a basically professional orientation, but which has also promoted the publication of several scientific journals. For many years the main theme of the APICS was the MRP, but lately it focused on the SC to the point that in 2014 it merged with the *Supply Chain Council.*

Between 1955 and 1960 Holt, Modigliani, Muth and Simon published their work on aggregate planning, with the first mathematical models for dealing with this problem.

Jay Wright Forrester (born 1918), was the founder of system dynamics (the first article on the subject was published in 1958). In his book *Industrial dynamics* (1961) he applies system dynamics to a production and distribution network and shows the effects that a disturbance on one element of the network has on the others. This is an analysis of the behaviour of a supply chain although the term did not yet exist. The analysis shows, by the way, that forecast errors, delays in the transmission of information and transport times combine to produce what was much later to be known as the *bullwhip effect* and properly speaking it is called the Forrester effect: a disturbance on one element of the network is amplified when transmitting upstream.

The *International Federation of Operational Research Societies* (IFORS) was formally constituted in 1959, although its first meeting was held in Oxford in 1957.

The changes in the way we view and deal with the problems of production management were reflected in the textbooks of Ellwood S. Buffa (the first, *Modern production management*, from 1961), with the quantitative approach that has dominated since then.

The appearance of the first PCs, in 1981, was very important, because it extended and facilitated access to hitherto centralised calculation resources.

The term *supply chain* (SC) appeared for the first time, without an explicit definition, in an article by Oliver and Webber (1982).

Shortly afterwards the idea grew that perhaps the use of quantitative techniques had gone too far, in the sense that attention had focused on the techniques themselves and not on the problems that they had to help solve. *The European Operations Management Association* (EUROMA, 1984) and the *Production and Operations Management Society* (POMS, 1989) also emerged in this period.

Finally, around the turn of the 21st century, coinciding with the growing availability and ease of data processing, talk began of *big data* and *analytics* (see Holsapple *et al.*, 2014)[15] both now in vogue. These terms are still a little nebulous, for which few matching definitions can be found, ranging from those that say that analytics is the technique of finding meaningful patterns in data, to those that consider it involves applying computer science, research and statistics to solving the company's problems.

2.2. Development of points of view and of terminology

Essentially, the terminological development has moved on from talking about production, to logistics and, finally, the supply chain (SC) and supply chain management (SCM).

---

[15] Davenport (2006) introduced the term. There is now has a certain perspective on the achievements, on the one hand, and the difficulties of applying analytics, on the other, (Cummings, 2016: "After more than 10 years it is still with us but if it is to survive it will need to become more rigorous, more scientific and more accountable"). See http://sloanreview.mit.edu/projects/strategy-drives-digital-transformation/.

Until well into the 20th century the talk was of production and manufacturing, with logistics being basically a military term. *Production management*, by the 1960s, had gradually been replaced by *production/operations management*, understood to refer to manufacturing and to services, although eventually *operations management* (concisely, OM) was to dominate.

Later, the term *logistics* emerged, and in the 1990s it was present in the academic field, all over the world. Outside universities, it was also of common used, but in a different sense.

As has been pointed out before, logistics is a term of military origin, which acquired a business meaning in the 1980s. Consequently, the term *industrial logistics* was sometimes used, which could be defined, for example, as the complete study of the movement of materials in industry and which includes the raw materials from source, to the delivery of the finished product to the market or customer. Therefore, this definition of industrial logistics includes manufacturing operations as part of the logistics.

Behind the academic use of the term logistics lays the idea of not considering the various parts of the supply-manufacturing-distribution flow in isolation, with the aim being to optimise the design and the operation of the system rather than those of each of its components.

Although there was an academic definition of logistics, it was interpreted very differently in business environments, where logistics refers to warehouses and transport and where a distinction is made between supply and distribution logistics. Here the persons responsible for logistics usually deal with one or the other and have, therefore, a responsibility for a part of the company's operations but not for harmonising all of them.

Perhaps this academic failure in relation to the integrating concept of logistics explains in good part why it has been replaced by that of supply chain, which encompasses and exceeds it. In the first years that followed its coining, the term *supply chain* had little impact, but afterwards it grew in popularity, first as an alternative to logistics and then as an expression of a new concept. It is significant that as part of the collection *Handbooks in Operations Research and Management Science,* by North-Holland, published in 1993, *Logistics of Production and Inventory* (Graves *et al.*, 1993) did not mention SC in the subject index; ten years later, *Supply Chain Management: Design, Coordination and Operation* (de Kok & Graves, 2003) appeared. In 2001 there was a lively controversy between logistics and SCM (Hall and Braithwaite, 2001; Lambert, 2001) and it still resonated in 2007 on the website of the *Council of Supply Chain Management Professionals* (which three years previously was called *Council of Logistics Management*), where, in spite of the extensive definitions of logistics and SC and the considerations about the border between the one and the other, things still remained unclear.

Even now there is no generally accepted definition of SC, although many coincide and, while there is no unanimity, the majority opinion is that SC is a new and more extensive concept than that of logistics, which reflects the new realities in organisations and helps in understanding the new problems that have arisen. A more detailed discussion can be found in Corominas (2013).

An SC refers to a product or a set of products and a dominant or leading entity. This entity decides the objectives, the criteria for evaluating the behaviour and configuration of the SC and establishes the main rules that govern the operation.

So, the SC can be defined as a set of entities that, as a result of a decision of the SC leader, collaborate in order to obtain, deliver, use, maintain and maybe recover a product or set of products. Therefore, the SC management refers to the flow (of people, materials, information, services, finances) between the entities that belong to the SC and the operations that should take place in some of them.

According to this definition:

- Only the entities that collaborate "as a result of a decision of the SC leader" have the status of SC members. Otherwise, even the simplest SCs would have exaggerated dimensions because the suppliers also have suppliers, and so on. Of course, the company that leads the SC decides who its suppliers are, but probably cannot decide who are the suppliers of its suppliers; in this case, the suppliers' suppliers should be considered external to the SC, even though they really do contribute.

- An SC is a network, that is, a set of elements with a set of relationships between them (by the way, an SC, in general, is not a *chain*).

- The SC not only includes the manufacturer (if this is the case) and its suppliers, but also the carriers, warehouses, retailers and even the customers themselves (Chopra and Meindl, 2016); after-sales services are also part of the SC.

- SC management (SCM) refers to all the activities taking place (among others, product design, marketing, operations, distribution, finance, and customer service), including the direct and reverse flows of all types.

The history of the discipline and the parallel terminological development show a progressive shift in points of view, which leads to an extension of the area under study as well as the research that has moved from the workplace to the factory, the logistics and production system, the supply chain, the environmental impact and sustainability. As Sarkis and Zhu (2017) say, "recent work shows a broader perspective on production research expanding to inter-organisational, supply chain, research". The same authors cite a premonitory snippet by Saunders (1971): "Historically, the production engineers and industrial engineers have stood at the periphery of the manufacturing plants and looked inward. They have confined their systems' viewpoint to what can be seen and controlled inside the four walls. This is no longer adequate. Instead of standing on the four walls and looking inward we must get to the periphery of our community ... and look at our total ecosystem where the manufacturing plant is merely one component of the system."

CHAPTER 3

THE FRAMEWORK OF THE SCIENTIFIC COMMUNITY: SOCIETIES, CONFERENCES AND JOURNALS

3.1. Scientific societies

The scientific societies[16], usually constituted as associations, are entities with a specific area of knowledge, offering their members a variety of support services for the development of their scientific activities in that field. They may have a strictly scientific character or be more orientated towards professional practice. There are normally no membership requirements and, in return for the membership fees, they provide:

- News about scientific developments, trends and activities, such as lectures, courses and conferences.
- Information about products (basically, publications and software) relevant to the research or to the professional activity.

[16] They should not be confused with the professional associations, with very different compositions and purposes to those of scientific societies.

- Courses organised by the scientific society itself.
- Publications.
- Conferences organised by the society.
- Discounts on fees for conferences organised by the society and subscriptions to publications.

Belonging to a scientific society does not imply any merit, but if you want to devote yourself to research and keep abreast of what is happening in your field you must be a member of one or more scientific societies. In other words, the fact of not belonging to any scientific society is an indicator of low involvement in research.

In Spain there are two associations that deal with the SC: ADINGOR (*Associación para el Desarrollo de la Ingeniería de Organización*, http://www.adingor.es/), founded in 1999, which organises an annual conference (IOC: *Congreso de Ingeniería of Organización*), and the SEIO (*Sociedad de Estadística e Investigación Operativa*, http://www.seio.es/), founded in 1962 as the *Sociedad Española de Investigación Operativa*, which organises sesquiannual conferences. The latter, however, has focused increasingly on operational research and statistics and the issues relating to the SC have little presence in their activities. The events organised by the technological platform Logistop (http://www.logistop.org/) are also interesting.

As has been indicated above, the origins of the United Kingdom's *Operational Research Society* (ORS, http://www.theorsociety.com/), chronologically the world's first in the field of operational research, date back to 1948 and their activities still reflect the multidisciplinary and application-oriented characters of the pioneer teams of operational research, which emerged in the UK. The ORS organises the YOR (*Young Operational Research*, biennial, 2017: YOR20) and OR (*Operational Research*, annual, 2017: OR59) conferences and publishes several journals (*Journal of the Operational Research Society* – initially, *Operational Research Quarterly* –, since 1950; *Journal of Simulation, OR Insight, Impact Journal* – the last two of a more informative nature – and others).

In the rest of Europe each country has its society of operational research or similar (in France, for example, the ROADEF: *Recherche Opérationelle et Aide à la Décision Française*, http://www.roadef.org/content/index.htm) integrated in the EURO (https://www.euro-online.org/web/pages/1/home), which organises two conferences every three years (the years when there is no IFORS conference, which takes place every three years) and sponsors the *European Journal of Operational Research* that, since it was first published in 1977, has achieved renown and prestige. EUROMA is the *European Operations Management Association* (http://www.euroma-online.org).

In other geographical areas there are similar organisations to the EURO (ALIO: Asociación Latino – Iberoamericana de Investigación Operativa: http://www-2.dc.uba.ar/alio//).

The *Institute of Industrial and Systems Engineers* (IISE, http://www.iienet2.org/) is based in the United States and publishes the monthly *ISE*, with a professional and informative focus, and is also responsible for IISE *Transactions*; INFORMS (https://www.informs.org/) organises an annual conference and has published *Operations Research* since 1952, *Management Science* since 1954, *Interfaces*, *OR/MS Today*, with a professional and informative orientation, as well as other journals; the *Production and Operations Management Society* (POMS) (http://www.poms.org/) organises an annual conference and has published *Production and Operations Management* since 1992; and the *American Production and Inventory Control Society* (APICS) (http://www.apics.org/), which sponsors the *Production and Inventory Management Journal*,with a practical orientation, and the *Journal of Operations Management*, which is now one of the most prestigious journals in this field.

The *International Federation of Operational Research Societies* (IFORS), (http://ifors.org/web/), organises a triennial conference and sponsors the journals *International Transactions in Operational Research* and *International Abstracts in Operational Research*.

The *International Foundation for Production Research* (IFPR) (http://www.ifpr-icpr.net/) was founded in 1971 in order to organise conferences that would provide contributions to the International *Journal of Production Research*, which began publication in 1961.

3.2. Conferences

In our field, a conference is a meeting at which the majority of attendees (from some hundred to a few thousand) present a synthesis of their most recent research work, finished or in progress. Workshops are a variant with a restricted subject matter and, therefore, with a smaller number of attendees.

So, our conferences are very different from the huge media events, similar in some ways to trade fairs or trade shows in which the majority of attendees are there to receive information about companies and products.

It is common for a conference to include plenary or semi-plenary sessions, in addition to parallel sessions (in the bigger conferences there may be dozens of simultaneous sessions). The plenaries include the formal opening and closing ceremonies and also lectures, on potentially interesting topics for all the attendees, run by people with recognised competence. The semi-plenaries (two or sometimes more simultaneously, although they are still called semi-plenaries) consist of lectures on specific topics. Parallel sessions include a certain number of contributions, with usually around fifteen or twenty minutes to present them, some time being reserved for the attendees questions.

With regard to what must be presented in order for a contribution to be accepted in a conference, there is a wide spectrum of possibilities. At one end, such as the IFORS, EURO and INFORMS conferences, only a very brief abstract (such as fifty words) is needed. This system allows assessing if the work is apposite or it is not, but does not, of course, guarantee its quality. Another option is to ask for a so-called extended abstract of, say, two pages. In some cases, the positive evaluation of the expanded abstract is only one requirement for presenting a relatively extensive text (possibly not as much as an average-length journal article), which will also be the subject of evaluation. In some cases these assessments are merely adjudicative (accepted or rejected); in others, they give indications for improvement, which can be incorporated in the final text.

Currently, the texts presented, whether they are abstracts, extended abstracts or full texts, are published by the conference in one or another support. Nowadays they can be on a pen drive, with documents in pdf, or paper. In the latter case it may be the proceedings of the conference or the publication of a selection of papers in a special issue of a journal or book.

The possibility of having work published is intended to be an inducement to participate in the conference, but it is ambivalent. The review processes of presentations at a conference are, for various reasons, among which the lack of time stands out, less demanding than those of the articles featured in a journal. As a result, publication of the work in the proceedings, or as a chapter of a book, is valued far less than the publication of an article in a recognised journal. On the other hand, a journal article should make an unprecedented contribution, so the publication of research results via the conference can prevent them being published as a journal article[17]. The result of all this may be that the conferences that

[17] In Chapter 4 (Ben R. Martin, "Ethics and integrity in publishing"), p. 36, Clark *et al.* (2016) there are some brief guidelines on this issue.

publish the complete texts only present results which are not important enough to be published in a journal.

The conferences are organised by scientific societies in which academic staff predominate, and therefore they also predominate among the conference attendees. Conferences are an important source of funding for these societies, since the participation fees often more than cover the costs.

Therefore, attendance at a conference is expensive, because in addition to the fee it is necessary to cope with the travel expenses and accommodation. Additionally, not much weight is attributed to the presentation of a paper in a conference when it comes to evaluating a curriculum.

So, what is the point in attending?

On the one hand, in the plenary and semi-plenary sessions sometimes emerging issues or an excellent synthesis of the state of research on a given topic may be presented. The parallel sessions provide access to the most recent results that, if they are important enough, will be published in a journal, albeit a few years later, due to the time required for the review and editing process (which is described in point 3.3); that is to say, journals are an instrument for staying up-to-date, but with some years of delay.

On the other hand, a conference allows you to submit work for the consideration of people working on the same or related topics, who can provide criticism and suggestions.

A conference is an area for meeting and making connections with these people. These relationships facilitate mobility (stages in research centres other than your own) and participation in joint projects with members of various research centres.

Finally, despite the little weight given in research evaluation to the presentation of papers at conferences, it is still an important item. Particularly, by default; someone who does research usually attends conferences (at least one each year) and so, the fact that a person does not attend conferences suggests that s/he is not sufficiently committed to the world of research.

3.3. Journals

The important research results are published in recognised journals (a little later we will clarify, or try to, what a recognised journal is). Although these journals also publish less important results, as well as important results being published in less recognised journals, it is generally agreed that quality research is published in these journals and that you cannot have a reputation as a researcher if you do not publish in these journals..

In the field of the SC, as in almost every other, there are thousands of people all over the world who research and who want to publish the results of their studies. Although some journals annually reach a total of hundreds of articles that cover thousands of pages, the number of articles that aspire to be published is far superior to what the most prestigious journals can print. A situation that has several consequences. First of all, the rejection rate of articles for these journals is 80% or even higher than 90%. Secondly, journals that aspire to be recognised are emerging. Finally, in a few years there has been a proliferation of journals that accept and rapidly publish anything and charge for doing so; these journals do not offer any guarantee of quality and you should not fall into the trap of posting there anything, as this may even result in a considerable drop in your prestige.

In previous points the names of some of the oldest and most prestigious journals have already appeared. There are many more and everyone who researches is familiar with their titles[18], from which many include *Supply Chain*

---

[18] To the aforementioned you can add, without claiming to be exhaustive: *Annals of Operations Research, Computers and Industrial Engineering, Computers and Operations Research, International Journal of Production Economics, Journal of Heuristics, Journal of Scheduling* and *Omega.*

Among those whose titles include 'Logistics': *EURO Journal on Transportation and Logistics, International Journal of Applied Logistics, International Journal of Logistics Management, International Journal of Logistics Research and Applications, International Journal of Logistics Systems and Management, Journal of Business Logistics, Journal of Logistics, Maritime Economics & Logistics, Naval Research Logistics* and *The Asian Journal of Shipping and Logistics.*

And, finally, among those who include 'Supply Chain': *International Journal of Integrated Supply Chain Management, International Journal of Supply Chain Management, Journal of Operations and Supply Chain Management, the Journal of Supply Chain Management, Supply Chain Forum: An International Journal* and *Supply Chain Management: An International Journal.*

As you can see, the fact that several journals have very similar or even totally matching thematic areas has forced them to adopt complicated titles to distinguish themselves from each other.

or *Supply Chain Management* (quite often, these journals are older than the term *Supply Chain* itself, and have changed their titles as a result of the emergence of the SC).

As we have seen in previous sections, many journals have sprung from the initiative of a scientific society, but others have been personal proposals which have been accepted by a publisher. Currently, the process of publishing in journals is usually undertaken by specialised publishing houses (which are increasingly larger and less numerous; currently Elsevier, Springer, Taylor & Francis and Wiley publish the vast majority of journals in our scientific field). The scientific responsibility falls on one or several editors and, much more tenuously and indirectly, on editorial boards that go by a variety of names. Generally, editors and board members are academics from a variety of countries, although in American journals they belong overwhelmingly to universities in the United States.

In the past, journals were published on paper and those that had already existed at the time continue to do so. All, or at least all those that are linked to major publishing houses, are published also in *pdf* format on their web pages. To have the journal on paper, or have access to the *pdf* of all the articles, a subscription fee must be paid (there is also the possibility of buying the *pdf* of each article). However, for some time now, there are journals that are published exclusively via the web and with open access (some give the option of obtaining printed copies, which have to be paid for); the so-called traditional journals also support and foster open access publication, which promotes the dissemination of their content, but requiring payment of a not insignificant sum to the publisher.

In the research projects funded by the Spanish *Plan Nacional de Investigación* one requirement from a few years ago is that articles derived from the project (published willingly or by obligation in high impact journals) should be open access (otherwise they are not considered). Open access via the website of the journal has, as has been said, a significant cost, which means that a good part of the grant finally would end up in the publishing houses, and the more productive the project the greater this amount. For this reason, the establishment of open access via repositories such as *UPC Commons* (https://upcommons.upc.edu/) is permitted, but you have to be very careful to respect the specific conditions of the journal in which the article has been published. These may be more or less flexible with regard to the uploading of the final or almost definitive version, the one that has been accepted but has not yet undergone the final processing at the hands of the publisher's technical services[19].

If you have access, you can find all the issues of the publication on each journal's website, (the early years might not be available if the journal is very old), as well as the accepted articles that have not yet been incorporated into an issue of the journal. In some cases these articles may correspond to different stages of the production process, such as accepted manuscripts and corrected proofs.

All over the world hundreds of journals, of international character, are published in each scientific area. But not all are equally demanding and publication of an article in one journal or another is not valued in the same way.

The most valued journals are those that are indexed. This means that they can be found in some databases, for which they have to meet specific, more or less demanding, conditions for each database. These databases deserve a variety of considerations, but clearly the most important is the JCR (*Journal Citation Reports*, from Clarivate Analytics), which includes SCI (*Science Citation Index*) and SSCI (*Social Science Citation Index*), each of which is subdivided into thematic categories, usually containing from a few dozen to a few hundred journals. Among the databases there is also *Scopus*, from Elsevier.

Within each category, the journals can be sorted according to various criteria, of which the *impact factor* is often considered the most important. This is a ratio that is calculated and published annually, in which the numerator is the number of times that articles published in the journal have been cited in certain journals, in a specific time since the publication of the article, and the denominator is the number of these articles. In many research evaluation systems the journals of each category, ranked from highest to lowest impact factor, are classified into tertiles (or perhaps quartiles) and the short term rating of an article depends on the

---

[19] In fact, there is an open debate on the role of the publishers (the vast majority of the journals are published by private companies) in the process of disseminating research results, as this is supported mainly by public funds; the authors and reviewers provide texts and reports to the publishers for no financial reward and the production costs of the articles are currently very low as a result of the widespread use of word processing and other software. The main obstacle to change in the current system are the evaluation procedures of the research (projects, groups and individuals) that are based on the impact factors and the fact that the high impact factor journals belong, with a few exceptions, to the these private publishers.

tertile containing the journal in which it is published (some evaluation systems only consider papers in journals belonging to the first tertile). This is based on the double assumption that a journal is good if, and only if, it has a high impact factor and that if an article is published in a good journal, the article is good[20]. Admittedly, as the pressure to publish in high impact factor journals is strong, these journals may be more demanding with regard to the quality of what they accept and, on the other hand, authors send what they consider good articles to these journals. In any case, it is clear that the publication of a paper in a high-impact factor journal does not guarantee a significant repercussion in the scientific community. In the long term, a significant indicator of an article's impact is the number of other articles published in indexed journals that cite it; however, this indicator can only be considered in the medium and long term, because it is very difficult for an article to be cited a significant number of times in the early years after its publication.

Without a doubt, there is a relationship between the quality of a journal and its impact factors in some or other databases. However, it is not always the journals with the highest impact factors that are considered the best by researchers in the field, among other things because sometimes the impact factor is not stable and the position of the journal in the category changes substantially from one year to

---

[20] As I say a little further on, the only way to know for certain if an article is good is to read it, if you are competent in the subject. However, a lot of articles would have to be read, taking many hours, to evaluate a person or group. It is easier and more economical, but not more reliable, to base the assessment on the support in which the article has been published, and not on the article itself.

The growth of research activity around the world and the need to evaluate it, with various objectives, has resulted in a new discipline, Informetrics: https://en.wikipedia.org/wiki/Informetrics ("Not to be confused with Informatics. Informetrics is the study of quantitative aspects of information. This includes the production, dissemination, and use of all forms of information, regardless of its form or origin. Informetrics encompasses the following fields:

- **Scientometrics**, which studies quantitative aspects of science
- **Webometrics**, which studies quantitative aspects of the The World Wide Web
- **Cybermetrics**, which is similar to webometrics, but broadens its definition to include electronic resources
- **Bibliometrics**, which studies quantitative aspects of *recorded* information.")

Elsevier has produced the journal *Informetrics* since 2007.

It can be said that in a way informetrics is a metaresearch (that is to say, a research on the research), the usefulness of which is not clear for everybody.

another. In addition, not all articles published in good journals are good, nor are all good articles published in journals considered to be good; therefore, there is no direct and universal relationship between the quality of the journal (measured by its impact factor) and the quality of the articles that are published in it. At the same time, doubts have been voiced about the reliability of the impact factor calculations. Finally, some journals apply various forms of pressure on editors and authors for their works to include many references to articles published in the same journal or in a related journal from the same publisher, which may even lead to the journal's temporary expulsion from a database. In general, the journals seek to increase their impact factor by means of various measures, some of which are appropriate and others, not. (Martin, 2016).

Therefore, other indicators of a journal's quality or of someone's research work have been proposed.. What has recently had most success is the *h* index (Hirsch index[21]), which is the biggest *h* integer such that at least *h* papers have received *h* citations. The *i10-index* is the number of publications that have received at least 10 quotes. In order to properly interpret these indicators it is necessary to know which journals are considered valid sources of citations (the same person can have several *h* index values, depending on what calculation tool is used).

The indicators mentioned above, or any other of a similar style (that is to say, that will present a complex reality, such as someone's research work, as a single numeric value) do not have to be absolute. In other words, although one researcher's *h* index is higher than another's, it does not mean that the work of the former is more important than that of the latter. It must also be taken into account that, for example, you can have a high *h* index as a result of signing several dozen articles as co-author with dozens, or even hundreds, of other people (as happens in some scientific areas).

The indicators give an idea of the quantity and the quality of the research activity, but the only way to know for sure if an article is good or not is to read it, if you are competent in that area, or else seek the opinion of a competent and reliable person who has read it.

---

[21] Analogous to what Arthur Eddington proposed for evaluating cyclists.

CHAPTER 4

WHAT CONSTITUTES A PIECE OF RESEARCH (A THESIS OR AN ARTICLE)?

The results of research are generally available as theses and articles and, secondarily, as chapters of books oriented towards research[22], conference presentations and working papers.

There are many useful tools for conducting research and publishing the results (a citation manager helps to save a bit of time) or that are essential for certain types of research (reference searchers). Depending on the case, it is necessary to master certain techniques (programming) or know how to make use of software (solvers, simulation languages, statistical packages). But it should not be forgotten that all this is worthless if you do not have a research topic and if you have not achieved any result in relation to it. To write a thesis or an article the only thing strictly necessary (assuming sufficient knowledge of the pertinent

[22] The book chapters tend to have less dissemination than journal articles, except, perhaps, that the book has exceptional characteristics because of the prestige of the publisher, the people who edit it and the quality of the contributions.

language) is to have something to say. If I may state the obvious: If you do not have anything to say you cannot explain anything. And, previously, you should have decided upon a research topic; Silver (2009) says it well: "One cannot do research without having a topic!"

And how to find a research topic? By knowing a lot about a subject or working with someone who does[23,24]. And, on the basis of this in-depth knowledge of a topic, devoting hours to thinking intensively about the gaps that exist and how to fill them.

However, it is not uncommon for the research texts to ignore this basic premise (that is to say, that it should have a subject, a research question, a procedure for answering the question and a relevant result), or mention it only in passing. In contrast, they give extensive coverage to issues that, while they are also important and need to be known, are not paramount; examples include the correct choice of the journal in which it is hoped to publish the article, the appropriate way to react to negative comments during the evaluation process, or how to know if a journal might consider an article, already published as a working paper or in the proceedings of a conference, as self-plagiarism. Of the thirty-two chapters of Clark *et al.* (2016), only one (Michael A. Hitt, "Publishing in the top journals: the secrets for success", pp. 163-167) focuses on the *development of the capacity to excel*, which must be based, according to Baron and Henry (2010), on "intense, prolonged and highly focused efforts to improve one's current performance" (p. 163): "Everyone in our profession can publish their work in top scholarly journals if they develop the capacity to excel, to identify a valuable research question, build a strong theoretical framework and ensure methodological rigor in the design of their research and in the development of the manuscript" (p. 166);

---

[23] Silver (2009) specifies a list of sources of research topics: reading, going to lectures and conferences, making use of information systems on research, teaching, consultancy, interacting with colleagues.

[24] As has already been pointed out above, the possibility of carrying out research individually cannot be completely ruled out, but, at least in the field of the SC, it should be regarded as extremely remote; usually, the research is done in groups and is based on the research projects, which are funded with plans such as the Spanish *Plan Nacional de Investigación* or the framework programs of the European Union.

I disagree again with Booth *et al.* (2008) for his description of the search for a subject as an individual adventure inspired by the interests of the person who wants to do research.

certainly it is true that those who achieve this will be able to publish in the highest level journals, but it is not easy and requires a sustained combination of ability and willpower over time.

In the aforementioned Silver (2009) it is also recommended to be active in more than one research topic, in the same way that a company should have a portfolio of products that are at different stages of their life-cycles[25].

Whatever type of research you want to carry out you should be aware that (I am not saying that this is good – on the contrary),as Singhal and Singhal state (2012b), "Editors and proofreaders of the research community tend to accept articles that represent established points of view and reject those which present alternative perspectives". Additionally, fragmentary contributions are published more than syntheses, perhaps as a result of a process in which it is difficult to say whether partial contributions predominate because they are easier to publish or if more are published because they are easier to obtain and research teams prefer them.

In order to discuss what the subject of a thesis or a research paper may be[26], some comments should be made about the various types of research in our field. Or, better still, on the characteristics that may be present, to a greater or lesser extent, in each category.

Essentially, we can include every research project in a space of three dimensions:

- theory-practice (application)
- quantitative-qualitative
- empirical-abstract.

[25] Someone who wants research to be a significant component of their activity throughout their working life must develop appropriate strategies (Mike Wright, "Sustaining a publications career", chapter 5, Clark *et al.*, 2016) and this analogy with the product portfolio of a company can be useful, provided that you do not confuse the results of research with goods (such as might occur from a mere reading of Wright's chapter) and not to forget that the aim is to make relevant contributions to the collective heritage of knowledge.

[26] Currently, a thesis must generate articles, even before it is defended in front of the examining board, because otherwise, unless it has very special characteristics that justify it, it will be considered to be of poor quality. Moreover, many doctoral programs will not accept a thesis if you have not generated publications in specified quantities and qualities.

Regarding the first of these dimensions, it must be said that although it refers to theoretical research and applied research, research in this field must always have application as its, more or less immediate, goal. The theoretical research involves developing concepts or theories. Practical research develops tools that will solve, or contribute to solving, specific problems or involves the solution of specific problems through an original combination of known tools.

The qualitative-quantitative dimension does not require explanation.

However, the empirical-abstract contrast is more complex and the terminology does not help much in understanding it. Gupta *et al.* (2006), on empirical research, tend to identify, improperly, what it is not empirical with what is not practical. In fact, all the research in the field of the SC must refer to reality, which is known by means of observation and, therefore, all research in the field of the SC could be ultimately considered empirical. Now, you can conduct research on some aspect of the SC, such as the production lines, already known as a result of observations made in previous researches or make observations as part of the same investigation. In the latter case it can be said that it is empirical research. Otherwise, it is abstract research.

I will use this abstract-empirical third dimension in order to demonstrate what a research topic may be.

But, whatever the type of research carried out, or intended to be, you should be well aware that the word "topic" can refer to a very diverse range of issues. Someone may say, for example, without being untruthful, that the theme of their thesis is the design of production systems, the design of production lines, of production lines with accessibility windows or the solution of the AWALBP-L3 problem (which is one of the types of problems that appear in the design of production lines with accessibility windows). If AWALBP-L3 is the subject of the thesis (or research paper) then surely one thing or another will be said depending on the audience being addressed. If you only know that you want to do some work on the design of production systems, you are still a long way from being able to start doing research in that area[27].

---

[27] On how to define the research subject so that it has an appropriate extension: Eco (1977), although the field referred to is that of Humanities.

Defining precisely the object of research (which is often done through the definition of a "research question") is critical. If the subject area is unattainable, by being excessively broad, a significant contribution can hardly be made. If it is too restricted, the results will surely only be relevant to very small groups. Except in the case of very new problems, the relative abundance of literature on the problem is usually a good indicator of whether the definition of the problem deviates too much, due to excess or by default, from the desired scope.

4.1. Abstract research

It is characterised by the use of mathematical models and algorithms to solve them in order to identify, and quantify if necessary, the most appropriate decisions. Therefore, it has many connections with operational research. The research can be focused on the description and understanding of the problems in order to formalise them as a prior step to modelling them, on their modelling or on algorithms for finding feasible, satisfactory or optimal solutions (in this last case, the field connects with those of artificial intelligence and computer science).

In this kind of research, one can distinguish, at least, the following modalities:

- **Review (the name that is currently more used than the more or less equivalent *state of the art*).**

  This option is relevant for any type of research, in the field of industrial engineering or any other.

  It consists in a complete and systematic study (which must be justified in the article itself with a description of the searches carried out – databases and keyword combinations –) of all the contributions published on a certain topic, the analysis of which can reveal emerging trends and orientations on the lines of future research. This kind of research is very time consuming, because it can involve accessing hundreds of publications, and requires a good knowledge of the subject and a good capacity for analysis to discover the main lines addressed and anticipate and suggest the lines of the future. But it is very useful to the scientific community in the corresponding field and much appreciated by the journals, because a good review generates many citations and, therefore, affects the journal's impact factor favourably.

- **Identification and formal description, oriented towards modelling, of a new problem.**

In the research group of which I am a member[28] we have often practiced this method on problems such as the planning and programming of schedules with annualised working time or with time accounts, the RTVP (*Response Time Variation Problem*, a problem of regular sequences), the GRCALBP (*General Resource Constrained Assembly Line Balancing Problem*, the design and allocation of tasks in production lines, with alternative processes and limited availability of several resources), the scheduling of activities taking learning and forgetting into account, and the AWALBP (*Accessibility Windows Assembly Line Balancing Problem,* the design and allocation of tasks in an assembly line where the accessibility of the tasks at each station is limited to a "window"), among others.

Nielsen (2004) examines the difference between two styles of research: *problem-solver* and *problem-creator*. Researchers probably combine both styles in varying proportions. Nielsen believes, and it is clear that he is right, that the problem-solver style is practiced more often, perhaps because the results are easier to recognise. In other words, it is easier to publish a work that solves a known problem, in a more or less new manner, than one that describes a novel problem. However, in the long run, if the problem is important, dealing with it is more helpful than improving, perhaps marginally, the tools for solving a known problem (which is not necessarily important). Silver (2009) says: "The incremental search has many more chances of giving rise to one or more publications in a relatively short time. However, the added value of each of these will be almost certainly eclipsed by that of an initial publication in a new area". Booth *et al.* (2008) also agree that it is more important to find a new problem to solve than one that has already been formulated.

- **Modelling a problem and model resolution.**

This is, first, the mathematical formulation of the objective function or functions and the constraints. Once the formal description of the problem has been made, modelling it is usually immediate for a competent specialist.

---

[28] Industrial Engineering and Logistics (EOLI), UPC (http://eoli.upc.edu/, https://www.ioc.upc.edu/investigacion/grupos/eoli, http://eoli.upc.edu/?set_language=en.

But the modelling itself is of very little or no interest if it does not serve to solve the problem. Therefore, it is not only a question of modelling, but of building models that can be solved within a time reasonable for the problem in question, with the tools available or that can be created. And this may present more difficulties, from the point of view of modelling. In any case, the model cannot be accepted as good unless computational experiments are carried out to verify that it can be solved, for dimensions that in practice are sufficient, in all plausible combinations of the data values. The design of data sets to test the model-resolution procedure combination is also a component of the research, as is the analysis of the results, which should normally be made from the computational point of view (time of resolution; bounds on the discrepancy between the optimal value and the value retrieved when the optimal solution cannot be found, etc.); it should also consider the characteristics of the solutions obtained, with the aim of detecting rules, general properties or generating *managerial insights*, that is to say, a better understanding of the problem studied from which guidelines for improving management can be derived.

With regard to this kind of research, we must distinguish between two types of models, which we may call the operational and conceptual, respectively.

The aim of the first is to give quantified indications about the decisions to be taken in a given context. Therefore, they must be based on realistic enough assumptions and require sufficiently precise data.

On the other hand, the purpose of the conceptual models is to demonstrate the most relevant variables for the behaviour of a system and how they can influence it. To achieve this, it is often necessary to make unrealistic assumptions, which simplify the problem and consequently make it possible to reach general conclusions. The EOQ formula is an example: the case of uniform demand made it possible, more than one hundred years ago, to obtain simple formulas that reflect the influence of the setup and holding costs and the amount of the demand on the optimal dimension of the orders and the cost of inventory management.

- **Identification and classification of the variants of a problem.**

There are problems (queues, scheduling, assembly line balancing or flexible organisation of working time, for example) that have numerous variants.

In this case it is very unlikely that all of them can be treated with the same instruments. So, identifying and classifying them favours a solution and facilitates communication between the research groups, as well as the organisation of the research. For some problems (at least, in the first three of the four mentioned at the beginning of this paragraph) encoding systems have been proposed to make the identification of variants more concise and accurate (with the added implication that, once the coding system is established, it is easy, with a simple exercise in combinatorics, to generate even thousands of variants, which do not necessarily exist in the real SCs).

- **Identification and demonstration of a property.**

For example, establishing that for a given type of problem all optimal solutions have a property *P* or that there is always an optimal solution that has the property *P*. In the first case we know that if a solution does not have the property it is not optimal and, therefore, we just study the solutions that have the property *P*. In the second, a solution that does not enjoy the property P may be optimal, but an optimal solution can be found although it is sought only among those solutions that have the property.

- **New design method.**

This type of research deals with establishing the steps to be taken when designing a system, the order in which they have to be performed, the inputs and outputs of each step, and to identify or define instruments to pass from the input to output efficiently. The best known example is, without a doubt, the SLP (Muther's *Systematic Layout Planning*); more recently, there is the SCOP method for the design of the SC (Corominas *et al.,* 2015).

- **New procedure[29] of resolution.**

This category includes many variants.

---

[29] I am referring to procedures and not algorithms, because the term procedure is broader and allows you to designate a combination of algorithms, or a meta-algorithm, such as metaheuristics, which are not algorithms but schemes from which an infinity of algorithms can be derived through the definition of partial procedures or fixing the values of parameters.

One of them is to design a procedure that includes several existing procedures as particular cases, by varying one of the routines or the values of one or more parameters. This is useful for teaching, because it allows students to learn more in less time and in a more structured way. As for research, it can open new possibilities, because the procedure that leads to integrating the already known can result in new procedures. Example: *branch-and-win* (Pastor and Corominas, 2004), which includes the variants of the *branch-and-bound* and other related procedures.

Another variant involves designing a new procedure, whether it is applicable to a broad set of problems (as in the case, for instance, of the simulated meta-heuristics annealing – SA: *simulated annealing* – or – TS: *taboo search* – useful for many problems of combinatorial optimisation) or to a specific problem (there are very many examples, since there are various procedures or algorithms for each of the many problems that have been studied and formalised). It is clear that designing a general procedure that shows significant progress is difficult and, therefore, is rare. However, the analysis of the literature on algorithms of combinatorial optimisation algorithms or for optimising multi-modal functions of real variables can give the feeling that important general procedures are found more often than not. Indeed, for some time now, new procedural proposals have been published based on analogies, sometimes merely presumed, with natural phenomena related to living or inanimate beings, but often the allegedly new procedure does not have a rational scheme that sustains it or is not nothing other than an already known procedure described with new terminology relating to the natural phenomenon on which it is supposedly inspired. This type of activity may flirt with the ridiculous (there is an imperialist algorithm, one of cockroach swarms, another of flying elephants, etc.) among other things, beyond the terminology, because the analogy is unsustainable or has no apparent connection with optimisation. For this reason, and because it can involve the discredit of really serious research, it has been severely criticised. This does not preclude it, for the time being, enjoying success, if this is measured by the number and dissemination of publications generated, but eventually seriousness and rigor can be expected to prevail. Here it is not possible to extend this debate more, but anyone wanting to do research in this mode must bear it in mind[30].

---

[30] In this regard, the basic reference is Sörensen (2015).

You can also include in this section the resolution of a specific problem with a procedure that has never been applied to this purpose beforehand. It also has an outlet, but sometimes the contribution is modest, and not very commendable. It will certainly be published if there is sufficient evidence or indications that the obtained results are better than those published previously (publication of a work that concludes that the algorithm proposed is not better than that already known[31] is extremely unusual, if not unheard of, even though it might be useful in saving a lot of research teams the task of repeating experiments that others have already done).

- **To deepen knowledge about the behavior of procedures, by themselves or in comparison with others (which includes identifying the most appropriate procedure, among those known, to solve a problem).**

The evaluation and comparison of algorithms, as well as the identification of the most appropriate algorithm to solve a given example of a problem[32], are activities for which there are practically no theoretical elements; therefore they are based almost exclusively on experimentation and the statistical analysis of the results. Consequently, they present theoretical and practical difficulties that are not generally solved and generate doubts about the strictly scientific character of these activities (Hooker, 1994, 1996), despite the publication of many articles that include them as essential content.

What Hooker means is that applying *m* algorithms to *n* different instances and elaborating tables of computing times and objective function values for each algorithm-instance pair cannot be considered a scientific activity; on the contrary, it would be a matter of explaining why one algorithm is

---

[31] In general, articles describing failures or bad practices are not published, even though, as Pfeffer says (2007), quoted in Singhal and Singhal (2012b) "knowing what does not work is often just as important as knowing what does work". In 2015, an attempt at crowdfunding to launch a journal that published negative results remained very far from achieving the goals proposed.

[32] A problem is a general question, such as finding a Hamiltonian circuit of minimum cost in a graph (TSP: Traveling Salesperson Problem). An instance of a problem is a particular case defined by data.

The relationship between problem and instance is analogous to that between a species of animal or plant and an individual of this species.

better than another or which algorithms are best for some examples and which for others.

It is clear that in order to decide the best algorithm it is necessary to define criteria. And to do so, in general, is not a trivial task and depends on the type of application. This is evident with regard to the computing time required for running the algorithm, which cannot be valued in the same way if it is for controlling a process in real time or for strategic capacity planning; on the other hand, it is important for real applications (in fact, extremely necessary in the vast majority of cases) to have the guarantee that the algorithm will provide a feasible solution of reasonable quality at a certain and determined time. The average quality of the solutions, measured with the value of the objective function, is an obvious and very general criterion; but very often an algorithm that provides high average quality solutions, but with a certain probability of very bad ones as well, will not be acceptable. In general, we want that an algorithm will, on average, provide good quality solutions and be robust, both with regard to the quality and the computing time.

The difficulties begin in the generation of data sets or sets of instances for testing. Are they representative? Do they contain examples of all of the types that may be important from the point of view of the behaviour of the algorithms? If a theory cannot be created and there are only the results of the computational experiment, the findings cannot be extrapolated to instances other than those included in the data sets. In addition, as the problems that are subjected to most research usually have reference data sets, accessible via the Internet, it turns out that the algorithms that successively overcome their predecessors are the algorithms that best solve, specifically, the examples of the reference data sets.

In addition, in a computational experiment it is not just the algorithms themselves that are compared, but, in the final instance, the codes, computer programs, each of which is one of the many possible ways to implement the algorithm operationally, executed on a specific computer. The fact that the code is accessible only to the group that created it, as often happens, has some important consequences:

– The experiment cannot be reproduced. If a proposal for an article presents a new algorithm and says that this has provided, in half the calculation time, solutions that improve the quality of results obtained with the current best-known algorithm by 20%, there is no possibility of confirming this when assessing the article.

- The result of the research is not directly transferable. If someone believes that the algorithm can be useful they must program it, possibly finding that their program is not as efficient as the original and, therefore, does not produce the expected results.

- Carrying out a computational experiment to compare algorithms requires a lot of work. A single person (or a team of people who work with very well-established and very strictly controlled methods) has to program all the algorithms to be compared and it has to be done, of course, in a way that could be described as neutral, that is to say which does not favor nor harm one or another algorithm (and how can you make sure of that?). An experiment of this type can give indications of fraud[33] if one of the algorithms compared shows a very different behaviour from that described when making it public[34].

Therefore, this type of research activity is very often based on some very weak premises. Since Hooker's article, little has progressed in solving the difficulties that he focused on. However, in recent years, Professor Smith-Miles (Smith-Miles *et al.*, 2014; Smith-Miles and Bowly, 2015) has been developing very promising methods and techniques for data generation and analysis of the results of computational experiments.

4.2. Empirical research

The empirical research is based on data obtained through observation of the reality of an organisation or set of organisations in order to:

---

[33] The number of fraud cases grows as the volume of research increases, in all areas of knowledge. For the motives behind fraud and the measures needed to prevent it, see: Akerlof and Shiller (2015).

[34] The concern for accuracy and reproducibility of computational experiments has intensified lately and has given rise to a very detailed and demanding proposal of protocols: *Good Laboratory Practice* (Kendall *et al.*, 2016).

- Provide elements to begin or advance the development of a theory.
- To verify or refute (*falsify*, according to Karl Popper)[35] a theory.

With regard to data collection there are several modalities that, for a particular study, can be used alone or combined[36]:

- **Qualitative.**

  The information comes from interviews and focus groups or similar techniques.

- **Case study[37].**

  A case study refers to a specific organisation or to a part of this organisation, the functioning of which is studied and analysed in depth, what requires a considerable time. There are also multicase studies, namely studies of a small number of cases that have some aspect in common (for example, companies of the same sector or with similar distribution systems).

  It is obvious that from one single case study one cannot draw out a theory, but it can give rise to the ideas needed to elaborate one. t is obvious that a theory cannot be drawn from one case study, but it can give rise to the

---

[35] Karl Popper stated that for a theory or proposition to be considered as scientific it had to be falsifiable, that is to say, it could be refuted by means of an observation. For example, the proposition "all beetles are black" would be refuted if a fuchsia coloured beetle were seen; on the other hand, it is not verifiable: the observation of millions of exclusively black beetles does not rule out that there may be other colours. It must be said, however, that the propositions "there are fuchsia beetles" or "not all beetles are black" are verifiable, but they are not falsifiable. "All people are mortal" is neither verifiable nor falsifiable. This text does not intend to enter into an analysis of Popperian epistemology (see Mosterín, 2013).

This brief discussion does highlight the importance of the definitions. There are propositions that are neither verifiable nor falsifiable because they are tautological: "all ravens are black" is a proposition of a different nature to "all beetles are black", because the Merriam-Webster dictionary defines ravens as "large, glossy-black birds" (which means that a bird with fuchsia-coloured feathers cannot be a raven), whereas the colour of beetles is not stated.

[36] At this point, I basically synthesise Gupta *et al.* (2006). See also Tharenou *et al.* (2007).

[37] In this regard, Voss *et al.* (2002), Barratt *et al.* (2011), Childe (2011) and Ketokivi and Choi (2014).

ideas needed to elaborate one. On the contrary, a single case study can lead to falsify (refute) a theory.

- **Field study.**

The data comes from visits to a number of organisations, but with fewer periods of interaction than in a case study.

- **Surveys (phone, mail, Internet).**

This is a widely used method, due to its low cost in comparison with the previous ones. The postal service, which was the largest channel a few years ago for some types of research, has been virtually abandoned in favour of the Internet, because of the great advantages in terms of convenience in filling in the questionnaire and the mailing and processing costs.

However, the difficulties presented by data collection using this method are considerable and often insurmountable. First, if the proportion of responses is not very high (which does not often happen), the results correspond to a sample that will inevitably be seen as biased, since it is reasonable to think that the organisations who have responded are somehow different to those that have not. In addition, in many cases, there is no control over who has responded, so the reliability of the information obtained is doubtful.

This does not mean, of course, that you cannot get valid conclusions with this method, as long as the necessary methodological precautions are taken.

- **Data from files.**

For many years there have been very important databases for research in the field of management. For example, the databases of company balance sheets allow you to study their financing structures and other aspects of their management.

However, there is currently a plethora of data available or potentially available, especially on people behaviour, thanks to the existence and widespread use of the Internet and also because companies, for a very low cost, can gather, store and analyse sales data, stocks, breakdowns, etc. This wide availability of data offers new possibilities for the use of quantitative models and has led to the rise of analysis and big data.

- **Laboratory experiments.**

  Some research into people's decisions in relation to consumption, risk or other subjects are based on data collected in laboratory experiments in which people (often students) assume the appropriate role under conditions that try to reproduce real-life. Since neither the conditions in the laboratory are real, nor the people who take the decisions are always those who take them in reality, the relevance of the results obtained is often questionable.

There is a variety of techniques for data analysis (descriptive statistics, multivariate analysis, structural equations and others), to which new ones are constantly added. Some of these techniques are very sophisticated and, as Guide and Ketokivi (2015) say, if not mastered perfectly they are at risk of being misused. Therefore it is necessary to ensure rigor or even add a specialised member to the research team because, as Guide and Ketokivi point out, people who carry out research in the field of OM (*Operations Management*), the area of the journal which they edit, are not (in general, of course) specialists in statistics[38].

[38] Guide and Ketokivi (2015) must be considered required reading for anyone who wants to do this kind of research.

CHAPTER 5

PUBLICATION OF RESEARCH RESULTS: STRUCTURE AND PROCESSING OF AN ARTICLE

5.1. The structure

The structure of an article, a chapter of a research book (very similar to an article), a working paper[39] or a thesis[40,41] is similar for research of the same type;

[39] A working paper is a text with characteristics similar or identical to those of an article, which is not published in a journal or a chapter of a book, but in a series of documents of a university department or institute, a research centre, a university or research institution's repository. This can be due to several causes: (i) accelerating the publication of research results; (ii) publishing a more extensive version (for example, with more details about the data and the results or demonstrations) that can fit into a journal; (iii) publishing a few results that may be of interest but without the sufficient quality level for acceptance in a journal.

[40] The extension of an article and a thesis are very different. A long article can occupy about fifteen pages of a journal, while a thesis normally contains up to several hundred pages (anyway, that of John Nash on non-cooperative games has only 32 typed pages typed, including the cover, two references, and the acknowledgments). A good thesis usually generates several articles and, on the other hand, everything that is said in a thesis is justified in much more detail than in an article.

[41] The following considerations are also applicable to oral presentations, such as those made at conferences, for example, with simplifications due to time limits.

but all depend, at least in the details, on the type of research in question. The structure of a review is, of course, different from that of an article that, for example, describes a new algorithm to solve a particular problem and its behaviour in relation to a set of instances.

The structure of an article can be as follows:

- Heading
- Abstract
- Keywords
- Introduction
- State of the art.
- Nucleus.
- Discussion, conclusions and perspectives.
- References.

The heading includes the title of the article (it is very important that it gives a precise idea of the content of the paper, because reading the titles of the articles is the first filter in discarding or retaining them for later consideration), the names of the authors, the institution or institutions to which they belong and their addresses. The corresponding author should also be identified, that is to say, the person that has taken charge of contact with the journal, from the submission of the original article to the correction of galley proofs, the same person to whom questions about the article should be addressed, if considered appropriate.

The order in which the authors appear is far from being a trivial issue. In the field of SC, as in many others, articles signed by only one person are scarce. When there are more authors[42], in the process of evaluating the research activity the question arises as to what contribution each has made, which, in general, is very difficult to ascertain (if only because in a group research paper it is difficult for

---

[42] In the *Times Higher Education* ranking of universities, articles signed by more than 1,000 people, classed as "freak research papers" are not taken into account (https://www.timeshighereducation.com/blog/world-university-rankings-blog-dealing-freak-research-papers%23survey-answer). In 2015 an article signed by 5,154 people was published, in which the content of the article made up only 9 of the 33 pages, the rest being dedicated to the list of signatories.

the members themselves to define their boundaries)[43]. There are teams that sign systematically in alphabetical order, but this has the drawback that the person listed first is often considered to be the one who has made the most important contribution (there are also those of the opinion that the most important contribution corresponds to the person signing in the last place) and, therefore, a person may be over-or under-valued as a researcher because their surname begins with one of the first or last letters of the alphabet. Signing with the same people is usually assessed negatively, because it is compatible with systematically assuming auxiliary tasks in the research team; therefore, sharing authorship with different groups of people in other articles is favourably considered, because it reflects well on the ability of the researcher to work in various environments.

The abstract sums up the content of the article in a nutshell (not much more than two hundred words in any case). This is the second filter for accessing the content and should be written with special care.

Keywords are words or groups of a few words that identify the subject area of the article (for example: *supply chain design, stochastic programming, matheuristics, manufacturing*). The number of keywords is variable; there are rarely fewer than three and not normally more than half a dozen, but in some cases there may be more than ten. Some journals have a list of keywords and specify that one or more of those that appear in the article must belong to the list. The choice of keywords is very important, because it conditions the selection of the people who will evaluate the article and is decisive with regard to finding it via search tools.

The introduction explains briefly (in maybe one or two pages) what the article is about and its contribution. The introduction is the third filter or, if you will, the third barrier prior to the reading of the full article. Currently, in our thematic scope, it is not uncommon for an introduction to start with a paragraph of rhetoric about globalisation[44], the growing demands of the markets in terms of quality and delivery times, etc. This kind of speech does not provide infor-

---

[43] On how to take each person's contribution to an article into account and the difficulties of doing so properly: de Mesnard (2017).

[44] Of course, in a scientific article you have to avoid explicit or implied value judgements (for example, by using the expression "developing countries") and maintain rigorous standards throughout the text. Right now (at the time of closing this text, in the forth quarter of 2017) it cannot be taken as read that globalisation is an unstoppable and increasingly intense process.

mation to any potential reader of the article and consumes a scarce resource, which is the space available. Therefore it is essential, from the very first paragraph, to explain clearly what the article is about and say, as soon as possible, what contribution it makes and why it should be considered important[45].

The state of the art or literature review is a more or less extensive review, according to the subject and the type of article, of its published antecedents. The aim is to describe what has been done worldwide in relation to the subject of the article and to show that the article's contribution, which will be described in more detail subsequently, is original.

The action taken and the results obtained are given in what I have called the nucleus (but this is never the title of a section of an article). The structure of this part of the article is quite varied, because it depends very much on the subject and how it has been dealt with. For example, if it refers to a new model for an optimisation problem, first the model will be presented (assumptions that define the problem, notation – data, parameters and variables –, formulation and explanation of the formulation) and then the computational experiment (the data sets used may be pre-existent or generated as part of the research paper itself: this last case must describe how they have been generated and include a link to make them accessible to other research teams; the results obtained and the analysis of the results).

The discussion may be more or less extensive (or even non-existent) depending on the nature of the article (for example: a review article requires an extensive discussion to extract the general ideas of the information analysed; an article that challenges commonly accepted propositions must discuss them in detail to be convincing). The conclusions contain an often very brief synthesis of the contributions of the article and, if applicable, the relevant managerial insights, that is to say, the ideas that emerge for management. It ends with indications about future lines of research related to the work that has been presented. In a review article these guidelines are particularly important; in fact they are the main contribution expected, as they have to guide the work of the research groups that operates in a specific area.

---

[45] Gerard P. Hodgkinson, in chapter 32 ("Publishing the interfaces of psychology and strategic management") by Clark *et al.* (2016) says (p. 283) that the authors of successful articles almost invariably make their intended contribution clear in the first three or four paragraphs (known as "the hook").

The references section contains the full identification of all publications (articles, books, chapters of books, web pages) mentioned in the text. The format is specific to each journal. The order is usually the alphabetical order of the surnames of the first signatories, but some journals use the order in which the publications are mentioned in the article.

The article may also include, although not always, a short section of credits and acknowledgements. Of course, mention must go to the research projects which have financed it totally or partially (this is a requirement associated with the granting of subsidies). Credit may also go to people who have contributed significantly, without being considered authors. It is also common to thank the editor and the reviewers (although these always remain anonymous) when their observations and advices contributed significantly to improving the quality of the article.

Sun and Linton (2014) set out the results of a comparative analysis between high-impact articles and rejections and concludes that the former devoted more space than the latter to the introduction and the conclusions, while in the rejections the proportion of text that explains what has been done and how is greater than in the high-impact articles.

Certainly, a common flaw even in published articles and even more in conference presentations is to explain the problem very quickly and then spend most of the time, or space, dealing with the technical details of the solution. But the precise and complete explanation[46] of the problem considered and justification as to why you want to solve it are fundamental. And one thing and the other will have to be said as soon as possible, whether in an article or in a presentation. Even when the problem is well known, good articles include a description.

Finally, it must be said that, after all these comments, the best way of knowing what a good research paper is like is to read research articles of quality.

---

[46] Before you start explaining how a problem has been solved, it must be fully defined, with all the suppositions that characterise it spelled out in a precise manner. Introducing the suppositions during the explanation of the solution process gives the impression that the suppositions were being added to avoid the difficulties as they occurred.

5.2. The process of publication

Once we have written an article (or, better, before you start to write[47]) we must decide what journal to send it to. In this regard, it is necessary to take into account:

- The orientation of the journal (quantitative, qualitative, empirical, focused on production, marketing or finance, etc.).
- Its standards of quality, which are closely correlated with the level of demand and the rate of rejection. You have to self-assess the quality of your work in order to estimate the level of the journals in which you can aspire to publish it.
- The information available about the terms of the reviewing and publication process. In some of these journals terms are brief and predictable, in others long and irregular. The brevity of time is always important, but particularly when you need the article to be published or, at least, be accepted as a guarantee of quality for a thesis or a merit for an accreditation process.

The article is usually posted via the Internet using specific but, of course, very similar applications for each publisher. Some of these applications limit the size of the article, in the number of words or characters and, in the event that the text exceeds the limit allowed, do not accept it.

Once received by the editor of the journal[48], the paper may be rejected immediately for several reasons:

---

[47] Reading a lot helps to write well. Troutt (1998) contains ideas and tips on improving the quality of the research and publication process. Also useful are books such as Booth *et al.* (2008). It is clear that a badly written article can harm a good research paper, but if the research is not good, a well written article will not save it.

[48] Lovejoy *et al.* (2011) describe briefly, but with sufficient detail, the process of review, acceptance or rejection of an article and give helpful hints on how to make a review when a journal asks us to.

- Inadequate language level (some journals allow you to resubmit on condition that this deficiency is corrected).

- Plagiarism and self-plagiarism. That is to say, significant fragments of the text coincide substantially with those of previously published articles, even by the same team or members of the same team. This is serious and may lead to a prohibition on sending articles to the journal for a long period, such as five years. In any case, plagiarism is obviously a practice to be abhorred and, if detected, a very considerable and hard to erase stain: achieving a good reputation is difficult, but not as hard as recovering it if it has been lost.

  Plagiarising is universally considered as one of the most serious ethical violations in research (which obviously also includes inventing or altering data or sources[49]), because it is an attempt to hijack, and without hardly any effort, results obtained by other people. Until recently, plagiarism could go unnoticed for a long time, or even forever. Now, however, there is software able to compare the proposed article with thousands of other publications and to detect matches, so you must take care in preparing the text so that a legitimate quotation, or the summary of a previous publication, is not mistaken for plagiarism.

- The inappropriate, in the editor's opinion[50], content of the article in relation to the orientation of the journal.

---

[49] Ethics in research is a topic that is increasingly becoming the order of the day. Unfortunately, it is because of the increasing number of cases in which it is known that it is not respected; Giménez-Toledo (2016) quotes a phrase from the book *Fakes & Frauds* (Myers, R., Harris, M., 1989), which should be borne in mind and is applicable to research and to the majority of human activities: "not all the individuals involved in the trade have been driven by the purest motives". There are some brief and interesting considerations on research and ethics in Booth *et al.*, (2008).

[50] I am referring to the editor, but it is also possible that these decisions are taken by an associate editor, responsible for a subject area. The editor assigns an *action editor* to each article, that is to say the person responsible for taking decisions about the article and communicating them to the correspondent author. This person may be the editor, an associated editor or, exceptionally, a member of the editorial board.

- Insufficient contribution or manifestly low quality, in the editor's opinion, in relation to the characteristics of the journal.

Once these filters have been overcome, the process of evaluation, known as peer review[51], begins. The editor proposes some experts in the subject to act as reviewers, to produce a report on the article. Their role is, on the one hand, to prevent the publication of error-ridden or very low quality articles. On the other hand they help to improve the quality of the articles that deserve, or might eventually deserve, publication.

The invitation to assess an article, obviously, can be accepted or not. In some subjects, only a very limited number of people are qualified to correctly evaluate an article and it may be the case that the editor might take months to get the required number of reviewers for the desired evaluation (two, in most journals; three in some others, to increase the probability of getting two reports completed in a reasonable time; and four, or even more, in some journals with very demanding standards). The increase in the number of items submitted to journals for consideration, even though the number of immediate rejections has also increased, means that the relative scarcity of trained and available reviewers is growing (and, according to the editors of Clark *et al.*, 2016, these people are increasingly less skilled and, therefore, the quality of their reports is becoming lower).

The evaluation may be blind (the reviewers see who the authors are, but the authors do not know who is reviewing them) or double blind (the authors

---

[51] The document *"Peer review. The nuts and bolts. A guide for early career researchers"* (http://www.senseaboutscience.org/data/files/resources/99/Peer-review_The-nuts-and-bolts.pdf) contains many interesting comments on the peer evaluation system and a discussion about its virtues and defects (in this regard, also Giménez-Toledo, 2016) and about the advantages and disadvantages of the blind and double blind systems.

"Peers" means that the people who evaluate have a similar level to those who have written the article. However, in practice, the reviewers wield great power in this role, because it is very difficult for an article to be accepted if a reviewer is strongly opposed (sometimes a wise editor may realise, especially if one or more positive reports have also been received, that a negative report is unfair or vague and accept the article, or keep it in the review process, despite the negative report; but it is difficult, particularly in prestigious journals that receive a lot of articles, because the editor, in order to maintain standards, has no other choice but to trust the reviewers' reports).

do not know who is reviewing them and the reviewers do not know who the authors are[52]).

The difficulty in finding reviewers is due not only to the low number of specialists available, but also to the characteristics of the work to be evaluated, which is quite difficult (reading thoroughly a paper, and writing and sending the report involves many hours), committed (the quality of a journal depends decisively on the quality of evaluation reports: see Söderlund and Bakker, 2014), and without any remuneration (by the way, the author of the scientific article is not paid either).

And for the system to work it has been estimated that each researcher must evaluate three articles for each one that s/he publishes.

Given the difficulty of finding reliable reviewers, the publishers have started to implement incentive schemes, such as honorific lists, diplomas for the best reviewers or discounts on books from the publishing house.

At the moment, these incentives are of little importance. People who agree to act as reviewers do so because it is a way to keep up to date in their field, it is seen as a necessary contribution to the scientific system and it forges a good relationship with the journal.

When a paper is submitted to a journal and shows enough evidence of quality, it must be evaluated within a reasonable time; frequently the publishers ask the proposed reviewer, if they have not accepted the task, to suggest the names of other reviewers. It is clear that this means that in some cases people who do not have the necessary knowledge nor experience end up evaluating an article and it can happen, as mentioned in Söderlund and Bakker (2014), that the author corrects the reviewer and not vice versa (remember that the current assessment

---

[52] Just because it is not seen does not mean it is not known. When it comes to a very specialised topic, the scientific community dealing with it is not very large, everyone knows each other and it is not hard to deduce the author from the text and references. Some journals allow the reviewer's identity to be revealed to the authors, but I think that this happens very rarely. Another thing is that the authors recognise a reviewer from what is said in the report, or because (unfortunately, as this is a reprehensible practice) they suggest, or even demand, that the authors include an excessive number of references corresponding to a particular author.

system is called peer review, in the sense of people of similar level; if the reviewer is not, by default, a *peer* of the author the logic of the system is upset and it does not work).

This phase of the process is the bottleneck and the results are sometimes unsatisfactory. Some reports show clearly that the reviewer has not understood the article fully or has not dedicated enough time and energy to it, what might be considered a perfunctory effort. Although they are not very common, articles of poor quality or containing errors are published, even in highly prestigious journals.

In the report the reviewer indicates the strengths and, above all, the weaknesses of the article, from the general approach to, perhaps, the typographical errors and makes a proposal or recommendation, with a list, which always includes:

- Accept as-is.
- Minor revision.
- Major revision.
- Reject.

The acceptance "as-is" for the original version of the article is extremely rare, even in cases of high-quality articles and recognised authors.

The boundaries between the lower and higher are not perfectly defined, but a minor revision only involves changes in some sections of the manuscript; when the reviewer makes this recommendation it usually means that acceptance will be recommended once the suggested changes have been made. A major revision involves restructuring and rewriting a significant part or even the entire article; the changes that have been introduced are very important and possibly definite only in a general sense; it cannot be assumed that, once revised, the article will be accepted (in fact, it is not uncommon for an article to be rejected after a major revision).

Some journals also include the option of rejection with the possibility of resubmitting (the difference with a major review is that, if the article is resubmitted, the process will start all over again).

The editor, after reading the reviewers' reports (which are sometimes contradictory) sends the corresponding author a decision letter, to which the reports are attached, indicating whether the article is to be accepted or rejected, etc. Some editors include their own evaluation of the reviewers' reports which indicate the importance the authors should give them (it could even be said that the editor indicates which reviews should be heeded and which not).

Typically, authors dislike the negative reviews of the reports, especially if they have been formulated in an undiplomatic manner. If the editor's decision is to reject the article it is inevitable that you will be in a bad mood (maybe even more so if it is a revised version than if it is the original). It may well be, of course, that the reviewers have not done their job well or that the editor, with two positive reports and one negative, has preferred to risk rejecting a good article rather than accept a bad one. There are specific examples of articles that have had a high impact but which were rejected by the first journal they were sent to[53].

But anyone who engages in research should know that these bad moments are inherent in the profession and, in any case, you have to analyse the reviewers' comments carefully and, as may be the case, the editor's, to see how they will help to improve the article. William H. Starbuck, in chapter 7 ("Squeezing lemons to make fresh lemonade: how to extract useful value from peer reviews") of Clark *et al.* (2016) proposed as a Golden Rule that "No reviewer is ever wrong!" and our experience has shown that it is almost always useful to bear this rule in mind and that it is easier to do so a few days after receiving the evaluation, if this is negative.

In general, people who have had a long and successful experience both as authors and in advising fledgling researchers (such as Clark *et al.*, 2016) recommend persistence and not losing heart when faced with a negative assessment or even a rejection of an article by a journal, because these people have also been through this experience, even when they already had a reputation. It is clear that the jour-

---

[53] The published opinions of authors on the reviews, and reviewers, of their articles are very varied.

For example, reading several chapters of Clark *et al.* (2016), which contain a good number of these opinions, brings to mind that everyone sees things from their own point of view. From authors who believe that it is not exceptional for editors and reviewers to take advantage of their positions of power to veto works because they contradict their own work, or because the article may anticipate their own work and diminish it, to others who appreciate the contributions of editors and reviewers because they believe that they have helped to significantly improve the quality of the article.

nals with most impact, with acceptance rates of 5% or 10%, reject quality articles (not all of the 90 or 95% that are sent and not published are bad articles: it is necessary to take into account, on the other hand, that there is already a prior filter on the part of the authors, that is to say, if they consider that an article is not up to standard they usually refrain from sending it to a very demanding journal, because the chances are that it is a waste of time).

If the editor's decision is to reject the article, you should consider whether to send it to another journal, after reviewing it and taking the comments received into account, or if it is better to archive it.

When the decision calls for a major review, the effort required for it, and the estimated odds of acceptance after review have to be weighed up, in order to conclude whether it is better to make the review, send the article to another journal, or archive it. When it comes to a minor revision, there is no doubt what you have to do, because it does not involve a lot of work and will quite likely lead to acceptance of the article.

In any case, in the revision you must respond to all comments received. One by one, (i) you have to take note and modify the article accordingly, (ii) argue that the comment is not right, or (iii) justify that what is proposed in the evaluation report is outside the boundaries of the article. What you should not do is ignore a comment, because this generates, logically, a very negative reaction on the part of the reviewer, since it implies a lack of consideration of their work.

Preparing a revised version of an article is a very demanding task, because usually preparing a research paper and writing the original version of the text requires complete dedication from the team. When, perhaps after a few months, the evaluations are received, the team, if it still remains as such, is working intensely on something else and dealing with the review makes it necessary to rearrange the work schedule. But you must not fall into the temptation and the error of responding superficially. It is necessary to modify the article and write to the editor and each of the reviewers in order to explain point by point what has been done, or why nothing has been done, in relation to the comments. And all this as soon as possible, if only to facilitate the reviewers' task, which is more cumbersome as more time passes between one review and another of the same article.

The recommendation to take all aspects of the evaluation reports into account and to do so in a way that facilitates the work of editors and reviewers does not imply, by far, that the authors have to adopt a servile attitude toward these persons. On the one hand, we must avoid flattering phrases such as "thank you for this particularly insightful comment" or similar. On the other hand, and this is most important, when you consider that a comment of the evaluation report is wrong it should be refuted, politely and with a solid argument to justify this position.

Once the revised version and the letters are prepared, they are sent to the journal for a new evaluation. This process can be repeated an indeterminate number of times. Revision should be considered as virtually inevitable, with two or three considered as normal. Furthermore, there are cases in which an article has been accepted after six reviews, but this is very exceptional.

If the article is finally accepted, it enters into the production process, which is usually relatively quick. The publisher sends the galley proofs for revision (for which it is usually given a very short time, such as forty-eight hours) or for some information to be added (such as a bibliographic reference), as well as the documents relating to copyright. Some journals publish the galley proofs on their websites without revision, with the indication that it is an accepted article and that the text has not been subjected to a final correction. Once the galley proofs have been checked, a few days will pass until the definitive text, with the DOI code, appears on the journal's website. Afterwards, months might pass until the article is included in an issue of the journal and assigned the corresponding page numbers.

Between one thing and the other, it is not common for the time between delivery and publication to be less than one year. Indeed, it is not unusual for it to take two or more years (up to six years, for example).

Therefore, the person who starts the research and has very limited time to present their theses or to establish their credentials has to choose a topic and type of article with a high probability of being accepted in a relatively short time frame. Later, with a consolidated history of research, they will be able to put the emphasis on importance, at the risk of facing longer publication deadlines.

In any case, due to the length of the evaluation and publishing process, which is largely unpredictable, even with a work of very high quality, in order to maintain a sustained pace of publication authors need to have articles at various stages of the process (if someone waits until the previous article is accepted before starting work on another, very long periods of inactivity will result, with little research and no publication).

Chapter 6

DIAGNOSES AND PERSPECTIVES

The early years of the 21st century provided some very favourable, interrelated, circumstances for analysing the past and present of the discipline in order to anticipate the future and help to prepare for it.

On the one hand, it had been about 50 years (the pioneering journals had already begun to celebrate their half-century) since the establishment of a paradigm that had started to be questioned, partly as a result of the evolution of supply, production and distribution systems (which led to the emergence and subsequent growth of the term *supply chain*) and in part by advances in calculation methods, techniques and instruments.

On the other hand, the institutionalisation of the OM was consolidated in the 1980s, separated from operational research: *Journal of Operations Management* (1980), EUROMA POMS, POM journal (see point 2.1).

In 2001, on the occasion of the 10th anniversary of the CIM Centre, I was invited to make a speech about research in the field of production management, which a few months later was repeated, with slight modifications in the Universidad Central de Las Villas, Santa Clara (Cuba). Among other things, I provided elements for distinguishing research from mere disclosure or charlatanism, analysed the divorce between reality and representation present in the academic world and criticised the unnecessary use of heuristics to solve problems that could be solved with exact procedures. Finally, I anticipated and advocated:

- Return to reality.
- Go beyond mere description.
- Do not give up on the exact solution of the problems susceptible to be modelled.
- Address new problems (new in either of two ways: (i) still unaddressed, such as many problems of strategic decision; (ii) that arise as a result of changes in reality, such as the logistics of e-commerce).
- Incorporate techniques and methods from other fields of knowledge[54].

Afterwards a couple of articles were published, from different perspectives and with specific contributions, coinciding in many important aspects with each other and with my point of view[55].

The first issue of volume 50 of the pioneer journal *Management Science* was dedicated to taking stock and defining the prospects of the various components of the thematic scope of the journal. It included Chopra *et al.* (2004), focusing on operations management (*"Five decades of operations management and the prospects ahead"*).

This work begins with a synthesis of the history of the discipline. It considers that its initial stages, in the 1960s, were characterised by the models and algorithms for solving tactical optimisation problems, with a single goal and a single decision-maker. In the decades of the seventies and eighties, the computing tools and the low availability of data were revealed as the bottlenecks for applications, some models were becoming stereotypes, many functional areas internalised optimisation and, while the academic world was navel-gazing, MRP, JIT and TQM had taken over in industry. Then, as a result of this situation, there was a reorien-

---

[54] Computer science, artificial intelligence, economics, game theory.

[55] Subsequently, Singhal and Singhal (2012a) also presented a synthesis of the development of the OM & SCM disciplines and a proposal of perspectives, but do not add substantial elements to those contained in the two previously mentioned articles.

tation of the research, which became more systematic and interdisciplinary and made use of more instruments, some imported, so to speak, from other disciplines; the authors proposed a return to the origins, that is to say, among other things, the contact with reality and interdisciplinarity. In 1987 the editorial policy of the Department of Manufacturing, Distribution and Service Operations of *Management Science* was also reformed: "of particular interest are papers that deal with strategic concerns such as the choice and impact of new production or information technology, and papers that may provide insight or simple models for guiding manufacturing or service policy". As for the future, the article anticipated that research would be more cross-functional, that is to say, they consider the functional areas of the company in a more integrated way and listed the seven following themes:

- SCs.
- OM- marketing interface.
- OM-finance interface.
- OM-organisations interface.
- Service operations.
- Operations strategy.
- Process design and improvements.

Two years later, the article "Supply chain management and Production and Operations Management: review, trends and opportunities" (Kouvelis *et al.*, 2006) appeared in volume 15 of *Production and Operations Management.* As the title shows, SC was the protagonist. First of all, the paper discusses the topics of all those articles relating to SC published in the journal, since its first issue:

- SC dynamics and the *bullwhip* effect.
- SC design, capacity and sourcing decisions.
- SC management practice: VMI (*Vendor Managed Inventory*) and re-engineering programs
- SC planning and scheduling.
- Teaching SC management.
- SC coordination: information sharing, incentives and contracts.
- Multi-channel coordination challenges: coordinating offline and online procurement and distribution.
- Design for SC management: postponement and product variety.
- Operational hedging and risk management in SCs.

Then, they indicated what they considered to be the emerging issues (very accurately, it can be said more than ten years later):

- SC disruptions of SC originated from man-made or natural.
- Closed-loop SCs.
- Environmental (green) aspects.

And, finally, what they considered to be opportunities for future research:

- "Packaging" the results from SCM research to SCM professionals.
- Empirically oriented research.
- Multiple agents.
- Future relationships.
- Services (including health and education).
- Interfaces with traditional OM topics (such as quality management).
- Impact of RFID (*radio frequency identification*).

Six years later the book *Switchpoints for the Future of Logistics* (Wieck *et al.*, 2012), was published, in which the switchpoints that are foreseen in the future of logistics and SC are listed. This is not a proposal of research lines, but of disjunctives, the result of which, now uncertain, will establish the research environment in the field of SC and the kinds of problems that will arise:

- Sources of raw materials; primary or secondary?
- Recycling: central or local?
- Food and water supply imbalances: aggravation or local solutions?
- Response to supply imbalance: large scale migrations or redistribution of resources?
- Fuel cost: extreme bottleneck or timely rescue?
- SCs: globalisation or regionalisation?
- Division of tasks: specialisation or integration?
- Life-cycle of products: increasingly shorter or stabilising?
- Business clusters: regional or sectorial?
- Shopping behaviour: home shopping or local retailing?
- Traffic flows: separation or integration?
- Standards for information technology: global standards or proprietary systems?
- Routing technology: local or central routing selection?
- Interaction at interfaces: face to face or automation?

- Last mile: bundle or separated networks?
- Level of sustainability: opportunity for distinction or homogenous standard?
- Driver for sustainability: demand or regulations?
- Environmental regulations: mandatory limits or financial instruments?

Currently, in my opinion (which in some respects is justified in Corominas, 2013) the important lines of research are those that relate to the following areas[56]:

- The, so to speak, traditional OM topics, that can be treated within or outside the framework of the supply chain:

  – Strategic planning.
  – Aggregate planning.
  – Supplier selection.
  – Stock management.
  – Scheduling.
  – Routes.

- Issues that arise in the context of the supply chain:

  – Terminology.
  – Design of the supply chain[57]:

    - Method.
    - Identification and evaluation of options.
    - Services.
    - Reliability, robustness.
    - Resilience.
    - Financial aspects.

  – Modelling tools.
  – Reverse logistics, closed-loop SCs.

[56] My opinion is that the indicated themes offer interesting perspectives for research. Not that these issues are the only ones on offer and, for sure, people with another field of expertise might indicate otherwise.

[57] See Corominas *et al.* (2015).

  - Sustainability, green logistics (in particular, the definition of criteria to take the three dimensions of sustainability into account: economic, environmental and social).
  - Humanitarian logistics.

- The impact (or lack of impact) of new realities in the design and management of the supply chain:

  - IoT (Internet of Things): impact on procurement, production and distribution.
  - Electronic sales.
  - Additive manufacturing (3D printing).

The problems associated with the design and the management of supply chains are extensive, complex and still largely unexplored. It is a perfect environment in which to conduct high impact research and, above all, relevant research, which ultimately is what matters, or should.

REFERENCES

Aghion, P., Dewatripont, M., Hoxby, C., Mas-Colell, A., & Sapir, A. (2008). *Higher aspirations: An agenda for reforming European universities.* Bruegel.

Akerlof, G.A., & Shiller, R.J. (2015). *Phishing for Phools: The Economics of Manipulation and Deception.* Princeton University Press. https://doi.org/10.1515/9781400873265

Baron, R.A., & Henry, R.A. (2010) How entrepreneurs acquire the capacity to excel: Insights from research on expert performance. *Strategic Entrepreneurship Journal,* 4, 49-65. https://doi.org/10.1002/sej.82

Barratt, M. Choi, T.Y., & Li, M. (2011). Qualitative case studies in operations management: Trends, research outcomes, and future research implications. *Journal of Operations Management,* 29, 329-342. https://doi.org/10.1016/j.jom.2010.06.002

Booth, W.C., Colomb, G.G., & Williams, J.M. (1995). *The Craft of Research.* The University of Chicago Press.

Childe, S.J. (2011). Case studies in operations management. *Production Planning & Control,* 22(2), 107-107. https://doi.org/10.1080/09537287.2011.554736

Chopra, S., Lovejoy, W., & Yano, C. (2004). Five decades of operations management and the prospects ahead. *Management Science,* 50(1), 8-14. https://doi.org/10.1287/mnsc.1030.0189

Chopra, S., & Meindl, P. (2016). *Supply Chain Management. Strategy, Planning and Operation.* 6th ed. Pearson, Prentice-Hall.

Clark, T., Wright, M., & Ketchen, D.J. Jr. (Eds.) (2016). *How to get published in the best management journals.* Edward Elgar. https://doi.org/10.4337/9781784714680

Corominas, A. (2013). Supply chains: What they are and the new problems they raise. *International Journal of Production Research,* 52(23-24), 6828-6835. https://doi.org/10.1080/00207543.2013.852700

Corominas, A., Mateo, M., Ribas, I., & Rubio, S. (2015). Methodological elements of supply chain design. *International Journal of Production Research,* 53(16), 5017-5030. DOI: 10.1080/00207543.2015.1013641. https://doi.org/10.1080/00207543.2015.1013641

Creedy, J. (2008). *Research without tears.* Edward Elgar.

Cummings, N. (2016). Has Analytics come of age? *Inside O.R.,* 549, p. 8.

Davenport, T.H. (2006). Competing on Analytics. *Harvard Business Review,* 84(1), 98-107

de Kok, A.G. & Graves, S.C. (2003). *Supply chain management: design, coordination and operation.* Handbooks in operations research and management science. North-Holland, 1993.

de Mesnard, L. (2017). Attributing credit to coauthors in academic publishing: The 1/n rule, parallelization and team bonuses. *European Journal of Operational Research,* 260, 778-788. https://doi.org/10.1016/j.ejor.2017.01.009

Eco, U. (1977). *Come si fa una tesi di laurea*[58]. Bompiani.

George Jr., C.S. (1974). *The history of management thought.* Prentice-Hall.

Giménez-Toledo, E. (2016). *Malestar. Los investigadores ante su evaluación.* Iberoamericana-Vervuert.

Graves, S.C., Rinnooy Kan, A.H.G., & Zipkin, P.H. (Eds.) (1993). *Logistics of production and inventory.* Handbooks in operations research and management science. North-Holland, 1993.

---

[58] There is a version in English: *How to Write a Thesis.* The MIT Press, 2015.

Guide, D., & Ketokivi, M. (2015). Notes from the Editors: Redefining some methodological criteria for the journal. *Journal of Operations Management,* 37, v-viii. https://doi.org/10.1016/S0272-6963(15)00056-X

Gupta, S., Verma, R., & Victorino, L. (2006). Empirical Research Published in Production and Operations Management (1992-2005): Trends and Future Research Directions. *Production and Operations Management,* 15(3), 432-448. https://doi.org/10.1111/j.1937-5956.2006.tb00256.x

Hall, D., & Braithwaite, A. (2001). The development of thinking in supply chain and logistics management. Cap. 6 de Brewer *et al., Handbook of logistics and supply-chain management,* 81-98. Elsevier.

Holsapple, C., Lee-Post, A., & Pakath, R. (2014). A Unified Foundation for Business Analytics. *Decision Support Systems,* 64, 130-141. https://doi.org/10.1016/j.dss.2014.05.013

Hooker, J.N. (1994). Needed: An empirical science of algorithms. *Operations Research,* 42, 201-212. https://doi.org/10.1287/opre.42.2.201

Hooker, J.N. (1996). Testing heuristics: We have it all wrong. *Journal of Heuristics,* 1, 33-42. https://doi.org/10.1007/BF02430364

Hutchins, R. (1959). *The University of Utopia.* The University of Chicago Press.

Kendall, G., Bai, R., Blazewicz, J., De Causmaeckers, P., Gendreau, M., John, R. *et al.* (2016). Good Laboratory Practice for optimization research. *Journal of the Operational Research Society,* 67, 676-689. https://doi.org/10.1057/jors.2015.77

Ketokivi, M., & Choi, T. (2014). Renaissance of case research as a scientific method. *Journal of Operations Management,* 32(5), 232-240. https://doi.org/10.1016/j.jom.2014.03.004

Kouvelis, P., Chambers, C., & Wang, H. (2006). Supply chain management and Production and Operations Management: review, trends and opportunities. *Production and Operations Management,* 15(3), 449-469. https://doi.org/10.1111/j.1937-5956.2006.tb00257.x

Lambert, D.M. (2001). The supply chain management and logistics controversy. Cap. 7 de Brewer *et al., Handbook of logistics and supply-chain management,* 99-126. Elsevier.

Lovejoy, T.I., Revenson, T.A., & France, C.R. (2011). Reviewing Manuscripts for Peer-Review Journals: A Primer for Novice and Seasoned Reviewers. *Annals of Behavioral Medicine,* 42, 1-13. https://doi.org/10.1007/s12160-011-9269-x

Martin, B.R. (2016). Editors' JIF-boosting stratagems – Which are appropriate and which not? *Research Policy,* 45, 1-7. https://doi.org/10.1016/j.respol.2015.09.001

Mosterín, J. (2013). *Ciencia, filosofía y racionalidad.* Gedisa.

Nielsen, M.A. (2004). *Principles of effective research.* http://michaelnielsen.org/blog/principles-of-effective-research/

Oliver, R.K., & Webber, M.D. (1982). *Supply-chain Management: Logistics Catches up with Strategy.* Outlook, Booz, Allen and Hamilton Inc. Reprinted 1992, in *Logistics: the Strategic Issues,* edited by M. Christopher, 63-75. Chapman Hall.

Pastor, R., & Corominas, A. (2004). Branch and win: OR Tree Search Algorithms for solving Combinatorial Optimisation Problems. *TOP,* 12(1), 169-191. https://doi.org/10.1007/bf02578930

Pfeffer, J. (2007). A modest proposal: how we might change the process and the product of managerial research. *Academy of Management Journal,* 50(6), 1334-1345. https://doi.org/10.5465/AMJ.2007.28166117

Sarkis, J., & Zhu, Q. (2017). Environmental sustainability and production: taking the road less travelled. *International Journal of Production Research,* 1-17. http://dx.doi.org/10.1080/00207543.2017.1365182

Saunders, B.W. (1971). Facilities Design: A Problem of System Analysis. *International Journal of Production Research,* 9(1), 3-10. https://doi.org/10.1080/00207547108929858

Silver, E.A. (2009). Some ideas on enhancing research productivity. *International Journal of Production Economics,* 118, 352-360. https://doi.org/10.1016/j.ijpe.2008.08.037

Singhal, K., & Singhal, J. (2012a). Imperatives of the science of operations and supply-chain management. *Journal of Operations Management,* 30, 237-244. https://doi.org/10.1016/j.jom.2011.11.003

Singhal, K., & Singhal, J. (2012b). Opportunities for developing the science of operations and supply-chain management. *Journal of Operations Management,* 30, 245-252. https://doi.org/10.1016/j.jom.2011.11.002

Smith-Miles, K., Baatar, D., Wreford, B., & Lewis, R. (2014). Towards objective measures of algorithm performance across instance space. *Computers & Operations Research,* 45, 12-24. https://doi.org/10.1016/j.cor.2013.11.015

Smith-Miles, K., & Bowly, S. (2015). Generating new test instances by evolving in instance space. *Computers & Operations Research,* 63, 102-113. https://doi.org/10.1016/j.cor.2015.04.022

Söderlund, J., & Bakker, R.M. (2014). The case for good reviewing. *International Journal of Project Management,* 32, 1-6. https://doi.org/10.1016/j.ijproman.2012.11.007

Sohdi, M.S., & Tang, C.S. (2014). Guiding the next generation of doctoral students in operations management. *International Journal of Production Economics,* 150, 28-36. https://doi.org/10.1016/j.ijpe.2013.11.016

Sörensen, K. (2015). Metaheuristics–the metaphor exposed. *International Transactions in Operational Research,* 22, 3-18. https://doi.org/10.1111/itor.12001

Sun, H., & Linton, J.D. (2014). Structuring papers for success: Making your paper more like a high impact publication than a desk reject. *Technovation,* 34(10), 571-573. https://doi.org/10.1016/j.technovation.2014.07.008

Tharenou, P., Donohue, R., & Cooper, B. (2007). *Management research methods.* Cambridge University Press. https://doi.org/10.1017/cbo9780511810527

Troutt, M.D. (1998). Some quality improvement principles useful in the OR/MS research and publication process. *Operations Research,* 46(2), 176-183. https://doi.org/10.1287/opre.46.2.176

Voss, C., Tsikriktsis, N., & Frohlich, M. (2002) Case research in operations management. *International Journal of Operations & Production Management,* 22(2), 195-219. https://doi.org/10.1108/01443570210414329

Wieck, I., Streichfuss, M. Klaas-Wissing, T., & Stölze, W. (2012). *Switchpoints for the Future of Logistics.* Springer. https://doi.org/10.1007/978-3-642-23493-4

AUTHOR

Albert Corominas is an emeritus professor of industrial engineering and operations research at the Universitat Politècnica de Catalunya (UPC), where he is a member of Institute of Industrial and Control Engineering, the Department of Business Administration, and the ETSEIB. He has degrees in computer science from the Universidad Politécnica de Madrid and in engineering from the Universidad de Valladolid (UVa), and a PhD from UVa. After having worked as a consultant, he has been teaching, for more than forty years, at the UPC and other universities. He has participated in numerous research projects and contracts. His research activities focus on modelling and solving, by means of optimisation techniques, industrial engineering problems and on the development of concepts and methodology for the design of supply chains. He is author or co-author of books and papers that have been published in AoOR, C&IE, C&OR, EJOR, IJFMS, IJPE, IJPR, INFOR, Interfaces, JoSch, JORS, Omega and ORL among other journals. Dr. Corominas is a member of the research groups EOLI (UPC), and of several scientific societies and editorial boards, including that of the International Journal of Production Research.

Made in United States
North Haven, CT
05 November 2026

21772138R00046